GCSE IN A WEEK

AUTHOR - FOWL

Use this day-by-day listin...

D1493622

Day		Page
1	Prime factors, HCF and LCM	2
1	Fractions	4
1	Percentages	6
1	Percentage change	8
1	Repeated percentage change	10
1	Reverse percentage problems	12
2	Ratio and proportion	14
2	Rounding and estimating	16
2	Indices	18
2	Standard index form	20
2	Formulae and expressions 1	22
2	Formulae and expressions 2	24
2	Brackets and factorisation	26
2	Solving linear equations	28
3	Solving quadratic equations	30
3	Simultaneous equations	32
3	Sequences	34
3	Inequalities	36
3	Straight-line graphs	38
3	Curved graphs	40
4	Distance–time graphs	42
4	Constructions	
4	Loci	

Day		Page
4	Angles	48
4	Bearings	50
4	Translations and reflections	52
5	Rotations and enlargements	54
5	Pythagoras' theorem	56
5	Similarity	58
5	Trigonometry	60
5	Trigonometry problems	62
6	Circle theorems	64
6	Measurement	66
6	Areas of plane shapes	68
6	Volumes of prisms	70
6	Dimensions/converting units	72
7	Pie charts	74
7	Scatter diagrams and correlation	76
7	Averages 1	78
7	Averages 2	80
7	Cumulative frequency graphs 1	82
7	Cumulative frequency graphs 2	84
7	Probability	86
	Tree diagrams	88
	Answers	90

T16896

The diagram below shows the prime factors of 60.

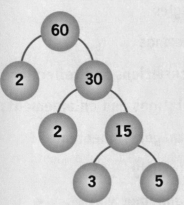

- Divide 60 by its first prime factor 2
- Divide 30 by its first prime factor 2
- Divide 15 by its first prime factor 3
- We can now stop because the second number 5 is prime.

As a product of its prime factors 60 may be written as:

$$60 = 2 \times 2 \times 3 \times 5$$

or

$$60 = 2^2 \times 3 \times 5$$

Highest Common Factor (HCF)

The largest factor that two numbers have in common is called the HCF.

Example

Find the HCF of 60 and 96.

- Write the numbers as the product of their prime factors.

$$60 = \boxed{2} \times \boxed{2} \qquad\quad \times \boxed{3} \times 5$$
$$96 = \boxed{2} \times \boxed{2} \times 2 \times 2 \times 2 \times \boxed{3}$$

- Ring the factors that are in common.
- These give the HCF = $2 \times 2 \times 3$

$$= 12$$

Any number can be written as a product of prime factors. This means the number is written using only prime numbers multiplied together.

Lowest Common Multiple (LCM)

This is the lowest number that is a multiple of two numbers.

Example

Find the LCM of 60 and 96.

- Write the numbers as products of their prime factors.

$60 = 2 \times 2 \times 3 \times 5$
$96 = 2 \times 2 \times 2 \times 2 \times 2 \times 3$

- 60 and 96 have a common prime factor of $2 \times 2 \times 3$, so it is only counted once.

- The LCM of 60 and 96 is

$2 \times 2 \times 2 \times 2 \times 2 \times 3 \times 5$

$= 480$

Progress check

1. Write these numbers as a product of prime factors:
   a) 50
   b) 360
   c) 16

2. Decide whether these statements are true or false.
   a) The HCF of 20 and 40 is 4.
   b) The LCM of 6 and 8 is 24.
   c) The HCF of 84 and 360 is 12.
   d) The LCM of 24 and 60 is 180.

3. Find the HCF and LCM of 36 and 48.

FRACTIONS

Addition and subtraction of fractions

Before adding and subtracting fractions, the denominators must be the same!

Example 1

$\frac{5}{9} + \frac{1}{7}$

$= \frac{35}{63} + \frac{9}{63}$ — Write both fractions with a demoninator of 63.
This is the lowest common multiple of 9 and 7.

$$\times 7 \qquad \qquad \times 9$$

$$\frac{5}{9} = \frac{35}{63} \qquad \qquad \frac{1}{7} = \frac{9}{63}$$

$$\times 7 \qquad \qquad \times 9$$

$= \frac{44}{63}$ — Now add the numerators but **not** the denominators.

Example 2

$\frac{4}{5} - \frac{1}{3}$ — Make the denominators the same: $\frac{4}{5} = \frac{12}{15}$, $\frac{1}{3} = \frac{5}{15}$

$= \frac{12}{15} - \frac{5}{15}$ — Replace the fractions with their equivalents.

$= \frac{7}{15}$ — Subtract the numerators but not the denominators.
The denominator stays the same.

Multiplication and division of fractions

When multiplying or dividing fractions, write out whole or mixed numbers as improper fractions before starting.

A fraction is part of a whole one. The top number is the numerator, while the bottom number is the denominator.

Example 1

$$\frac{2}{7} \times \frac{4}{5}$$

$$= \frac{2 \times 4}{7 \times 5}$$ Multiply the numerators.
Multiply the demoninators.

$$= \frac{8}{35}$$

Check whether the fraction simplifies.

Example 2

$$2\frac{1}{3} \div 1\frac{2}{7}$$

$$= \frac{7}{3} \div \frac{9}{7}$$ Convert to top heavy fractions.

$$= \frac{7}{3} \times \frac{7}{9}$$ Take the **reciprocal** of the second fraction and multiply.

Reciprocal means to turn the fraction upside down.

$$= \frac{49}{27}$$

$$= 1\frac{22}{27}$$ Rewrite as a mixed number.

Progress check

1 Match the answer to the question. The first one has been done for you.

a) $\frac{1}{3} + \frac{1}{2}$ $\frac{3}{55}$

b) $\frac{2}{5} - \frac{1}{7}$ $\frac{12}{35}$

c) $\frac{3}{5} \times \frac{1}{11}$ $\frac{5}{6}$

d) $\frac{4}{7} \div 1\frac{2}{3}$ $\frac{9}{35}$

2 Without using a calculator, work out the following:

a) $\frac{2}{3} + \frac{5}{7}$

b) $2\frac{1}{3} \times 4\frac{1}{2}$

c) $\frac{9}{7} \div \frac{3}{5}$

This is the percentage sign

%

Percentage of a quantity

OF means multiply.

Example 1

Find 30% of 80 kg.

$$\frac{30}{100} \times 80 = 24 \text{ kg}$$

 On the calculator, key in 30 ÷ 100 × 80 =

30% = 0.3 — this is known as the **multiplier**

For the **non-calculator** paper:

- find 10% by dividing by 10

  10% of 80 kg

  = 80 ÷ 10

  = 8 kg

- then multiply by 3 to get 30%

  3 × 8

  = 24 kg

Percentages are fractions with a denominator of 100.

Example 2

A CD player costs £65. In a sale it is reduced by 15%. Work out the cost of the CD player in the sale.

Method 1

15% of £65

$= \frac{15}{100} \times 65$

$= £9.75$

$= £65 - £9.75$

$= £55.25$ (price of CD player in the sale)

Method 2
(using a multiplier)

$1 - 0.15 = 0.85$
(0.85 is the multiplier)

0.85×65

$= £55.25$

⬭ One quantity as a percentage of another

To make the answer a percentage – multiply by 100%

Example 2

Matthew got 46 out of 75 in a Science test. What percentage did he get?

$\frac{46}{75} \times 100\%$ Make a fraction.

$= 61.\dot{3}\%$ Multiply by 100%.

⌨ On the calculator key in:

| 46 | ÷ | 75 | × | 100 | = |

DAY 1

Example

Tammy bought a flat for £185 000. Three years later she sold it for £242 000. What is her percentage profit?

Profit = £242 000 − £185 000

     = £57 000

Percentage profit = $\dfrac{57\,000}{185\,000} \times 100\%$

     = 30.8% (3 sf)

$$\text{Percentage change} = \frac{\text{change}}{\text{original}} \times 100\%$$

Percentage change might include increase, decrease, profit, loss, error etc.

Example

John bought a bike for £135. Six months later he sold the bike for £95. Work out his percentage loss.

Loss = £135 – £95

    = £40

Percentage loss = $\frac{40}{135} \times 100\%$

          = 29.6% (3sf)

Remember to divide by the original value.

Progress check

1 Imram bought a car for £8500. Two years later he sold it for £4105. Work out his percentage loss.

2 In a sale a coat is reduced from £135 to £72. Work out the percentage reduction.

3 The answers to the following questions are written on the cards below. They have all been rounded to 2sf.

| 25% | 13% | 12% | 26% | 19% |

Match the cards with the questions.

a) A sofa bought for £650 and sold for £820. Percentage profit?

b) A house bought for £95 000 and sold for £107 000. Percentage profit?

c) A car bought for £15 205 and sold for £12 370. Percentage loss?

d) A sweater bought for £65 and sold for £49. Percentage loss?

DAY 1 2 3 4 5 6 7

Repeated percentage change

A car was bought for £12 500. Each year it depreciated in value by 15%. What was the car worth after three years?

> **You must remember not to do: 3 × 15% = 45% reduction over 3 years!**

Method 1

- Find 100 − 15 = 85% of the value of the car first.

  Year 1 $\frac{85}{100} × £12\,500 = £10\,625$

- Then work out the value year by year. (£10 625 depreciates in value by 15%.)

  Year 2 $\frac{85}{100} × £10\,625 = £9031.25$

  (£9301.25 depreciates in value by 15%.)

  Year 3 $\frac{85}{100} × £9031.25 = £7676.56$

Method 2

- A quick way to work this out is by using a multiplier.

- Finding 85% of the value of the car is the same as multiplying by 0.85.

  Year 1: 0.85 × £12 500

  Year 2: 0.85 × £10 625

  Year 3: 0.85 × £9031.25

- This is the same as working out $(0.85)^3 × £12\,500 = £7676.56$

Compound interest is where the bank pays interest on the interest earned as well as on the original money.

○ Compound interest

Charlotte has £3200 in her savings account and compound interest is paid at 3.2% p.a. How much will she have in her account after four years?

$100 + 3.2 = 103.2\%$

$\qquad = 1.032$ This is the multiplier.

Year 1: $1.032 \times £3200 = £3302.40$

Year 2: $1.032 \times £3302.40 = £3408.08$

Year 3: $1.032 \times £3408.08 = £3517.14$

Year 4: $1.032 \times £3517.14 = £3629.68$

Total = £3629.68

A quicker way is to multiply £3200 by $(1.032)^4$

number of years

$£3200 \times (1.032)^4 = £3629.68$

original   multiplier

Progress check

1. Reece has £5200 in the bank. If compound interest is paid at 2% p.a., how much will he have in his account after 3 years?

2. Complete this statement:
   Some money is invested for 2 years. If compound interest is paid at 2.7% p.a. the multiplier would be

   ....

3. Mr Singh bought a flat for £85 000 in 1999. The flat rose in value by 12% in 2000 and 28% in 2001. How much was the flat worth at the end of 2001?

DAY 1

The price of a television is reduced by 20% in the sales.
It now costs £350. What was the original price?

- The sale price is 100% − 20% = 80% of the pre-sale price (x)

- 80% = 0.8   This is the multiplier.

- 0.8 × x = £350

$$x = \frac{£350}{0.8}$$

Original price = £437.50

Check

$$\times\ 0.8$$

original price $\longrightarrow$ new price

$$\div\ 0.8$$

Does the answer sound sensible?
Is the original price more than the sale price?

A telephone bill costs £169.20 including VAT at $17\frac{1}{2}$%.
What is the cost of the bill without the VAT?

- The telephone bill of £169.20 represents
  100% + 17.5% = 117.5% of the original bill (x).

- 117.5% = 1.175   This is the multiplier.

- 1.175 × x = £169.20

$$x = \frac{£169.20}{1.175}$$

Original bill = £144

Check

$$\times\ 1.175$$

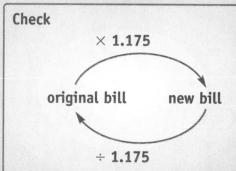

original bill   new bill

$$\div\ 1.175$$

These are when the original quantity is calculated. They are quite tricky so think carefully!

The price of a washing machine is reduced by 15% in the sales. It now costs £323. What was the original price?

- The sale price is 100% − 15% = 85% of the pre-sale price (x).

- 85% = 0.85   This is the multiplier.

- 0.85 × x = £323

$$x = \frac{£323}{0.85}$$

Original price = £380

Progress check

1 Each item listed below includes VAT at 17.5%. Work out the original price of the item.

 a)   a pair of shoes: £62

 b)   a coat: £125

 c)   a suit: £245

 d)   a TV: £525

2 In the winter sales the price of the items below are reduced by 15%, and the new prices are given. Joseph works out the original prices and writes them below. Decide whether Joseph is correct.

 a)   CD player

   £60

   original price:

   £70.59

 b)   Mountain bike

   £240

   original price:

   £276

 c)   Sweater

   £30

   original price:

   £35.29

DAY 1
2
3
4
5
6
7

Sharing a quantity in a given ratio

To share an amount into proportional parts, add up the individual parts and then divide the amount by this number to find one part.

Example

£155 is divided in the ratio of 2:3 between Daisy and Tom. How much does each receive?

$2 + 3 = 5$ parts ⟶ Add up the total parts.

5 parts = £155

1 part = £155 ÷ 5 ⟶ Work out what one part is worth.

= £31

So Daisy gets $2 \times £31 = £62$ and Tom gets $3 \times £31 = £93$.

Check: £62 + £93

= £155 ✔

Best buys

Use unit amounts to help you decide which is the better value for money.

Example

The same brand of breakfast cereal is sold in two different sized packets. Which packet represents the better value for money?

- Find the cost per gram for both boxes of cereal.

  125 g costs £1.65 so $165 ÷ 125 = 1.32$p per gram

  500 g costs £3.15 so $315 ÷ 500 = 0.63$p per gram

- Since the larger box costs less per gram, it is the better value for money.

Increasing and decreasing in a given ratio

A patio took 4 builders 6 days to build.

At the same rate how long would it take 6 builders?

Time for 4 builders = 6 days

Time for 1 builder = 6 × 4 = 24 days

Time for 6 builders = 24 ÷ 6 = 4 days

It takes 1 builder four times as long to build the patio.

Example

A photograph of length 12 cm is to be enlarged in the ratio 4:5.

What is the length of the enlarged photograph?

12 ÷ 4 = 3 cm	Divide 12 by 4 to get 1 part.
3 × 5 = 15 cm	Multiply this by 5 to get the length of the enlarged photograph.

In a survey, only 1 out of every 20 people did not answer the phone if it rang while they were watching TV.

Rubbish

However, 19 out of every 20 teenage boys ignored the phone if it rang while they watched TV.

That's more like it.

Decimal place

When rounding numbers to a given number of decimal places (dp) count the number of places after the decimal point, then look at the next digit after the one you want.

> **If the number is 5 or bigger, round up.**
>
> **If the number is 4 or smaller, the digit stays the same.**

$$2.3725 = 2.373 \text{ (3dp)}$$

The digit is 5 so round up the 2   The 2 rounds up to a 3

Significant figures

The first significant figure is the first digit that is not a zero.
The 2nd, 3rd... significant figures follow on after the first digit.
They may or may not be zeros.

$$6347 = 6350 \text{ to 3sf}$$

The digit is 5 or more,   You must fill in the end zeros.
so the 4 rounds to a 5   This is often forgotten.

> **The same rules apply as in decimal places.**

Rounding to a certain number of decimal places or significant figures is important in the exam. If you don't do this right, you lose the marks!

10 MINS

DAY 2

Estimating

When estimating the answer to a calculation, you must round the number to 1 significant figure.

$$\frac{273 \times 49}{28} \approx \frac{300 \times 50}{30} = 500$$

> $\approx$ **means approximately equal to**

If a measurement is accurate to some given amount, then the true value lies within half a unit of that amount.

Example

If the weight (w) of a cat is 8.3 kg to the nearest kilogram, then the weight would lie between 8.25 kg and 8.35 kg

$$8.25 \leqslant w < 8.35$$

lower bound   upper bound

Progress check

1. Put a ring around the correct answer.
   3724 rounded to 2 significant figures is:
   a) 3800
   b) 37
   c) 38
   d) 3700

2. Patrick made the following statements. Decide whether the statements are true or false.
   a) 4625 rounded to 3sf is 4630
   b) 2.795 rounded to 1dp is 2.7
   c) 0.00527 rounded to 2sf is 0.0053
   d) 37 062 has 4 significant figures

3. Work out an estimate for:
   $$\frac{2.09 \times 794.6}{0.48}$$

4. A piece of metal is 64 cm, correct to the nearest centimetre. Write down the minimum value that the length could be.

INDICES

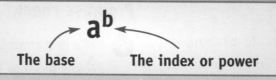

The base   **The index or power**

The **base** has to be the **same** when the rules of indices are applied. The laws of indices can be used for numbers or in algebra

Laws of indices

$$a^n \times a^m = a^{n+m}$$

$$a^n \div a^m = a^{n-m}$$

$$(a^n)^m = a^{n \times m}$$

$$a^0 = 1$$

$$a^1 = a$$

$$a^{-1} = \frac{1}{a^1}$$

Examples with numbers

1. Simplify the following, leaving your answer in index notation.

   a) $5^2 \times 5^3 = 5^{2+3} = 5^5$

   b) $8^{-5} \times 8^{12} = 8^{-5+12} = 8^7$

   c) $(2^3)^4 = 2^{3 \times 4} = 2^{12}$

2. Evaluate:

   | **Evaluate means to work out.** |

   a) $4^2 = 4 \times 4 = 16$

   b) $5^0 = 1$

   c) $3^{-2} = \frac{1}{3^2} = \frac{1}{9}$

10 MINS

Examples with algebra

1. Simplify the following:

a) $a^4 \times a^{-6} = a^{4-6} = a^{-2} = \dfrac{1}{a^2}$

b) $5y^2 \times 3y^6 = 15y^8$

The numbers are multiplied.

The indices are added.

c) $(4x^3)^2 = 16x^6$

Remember to square the 4 as well.

2. Simplify:

a) $\dfrac{15b^4 \times 3b^7}{5b^2} = \dfrac{45b^{11}}{5b^2} = 9b^9$

b) $\dfrac{16a^2b^4}{4ab^3} = 4ab$

Progress check

1 Simplify the following, leaving your answers in index form.

a) $6^3 \times 6^5$

b) $12^{10} \div 12^{-3}$

c) $7^{10} \div 7^{-14}$

d) $(5^2)^3$

2 Simplify the following:

a) $2b^4 \times 3b^6$

b) $8b^{-12} \div 4b^4$

c) $(3b^4)^2$

d) $\dfrac{9b^6 \times 2b^5}{3b^{-3}}$

DAY 2

DAY

1
3
4
5
6
7

STANDARD INDEX FORM

When written in standard form a number will be written as $a \times 10^n$

A number between 1 and 10
$1 \leq a < 10$

The value of n is the number of places the digits have moved to return the number to its original value.

Example

Write 2 730 000 in standard form.

- Move the decimal point to between the 2 and 7, to give 2.73 $(1 \leq 2.73 < 10)$

- Count how many spaces the decimal point has to move to restore to its original number

2 7 3 0 0 0 0 (6 places)

So $2\,730\,000 = 2.73 \times 10^6$

- If the number is big, n is positive.
- If the number is small, the n is negative.

Example

Write 0.000046 in standard form.

4.6×10^{-5}

- Put the decimal point between the 4 and 6.

- Move the decimal point back five places to restore the original number.

- The value of n is negative.

Standard index form is used to write very large or very small numbers in a simpler way.

 To put a number written in standard form into your calculator you use the [EXP] or [EE] key.

$(2 \times 10^3) \times (6 \times 10^7) = 1.2 \times 10^{11}$

This would be keyed in as

[2] [EXP] [3] [×] [6] [EXP] [7] [=]

On a non-calculator paper you can use indices to help you work out answers.

$2 \times 10^3 \times 6 \times 10^7$

$= 2 \times 6 \times 10^3 \times 10^7$

$= 12 \times 10^{3+7}$

$= 12 \times 10^{10}$

$= 1.2 \times 10 \times 10^{10}$

$= 1.2 \times 10^{11}$

Progress check

1 Write in standard form:
  a) 64 000
  b) 271 000
  c) 0.00046
  d) 0.000000074

2 Without a calculator, work out the following. Leave in standard form.
  a) $(3 \times 10^4) \times (4 \times 10^6)$
  b) $(6 \times 10^{-5}) \div (3 \times 10^{-4})$
  c) $(5 \times 10^6) \times (7 \times 10^9)$

3 Work these out on a calculator:
   a) $(4.6 \times 10^{12}) \div (3.2 \times 10^{-6})$
  b) $(7.4 \times 10^9)^2 + (4.1 \times 10^{11})$

DAY 2

DAY 2

- A term is a collection of numbers, letters and brackets, all multiplied together, e.g. $6a$, $2ab$, $3(x - 1)$

- Terms are separated by $+$ and $-$ signs. Each term has a $+$ or $-$ attached to the front of it.

$$5ab - 3c - 6b^2 + 7$$

invisible $+$ sign | ab term | c term | b^2 term | number term

- $3c$ means $3 \times c$ or $c \times 3$ or $c + c + c$

- ab means $a \times b$ or $b \times a$

- b^2 means b multiplied by itself $= b \times b$

> $3b^2$ means $3 \times b \times b$

- $a \div 2$ can be written as $\frac{a}{2}$

- $c \times a \times 5 = 5ac$ — the number usually comes first and then the letters in alphabetical order.

- $3a^2$ is not the same as $(3a)^2$
  $3a^2$ is 3 lots of just a^2
  $(3a)^2$ is 3 multiplied by a, then all of it squared.

> **Simplify means make the expression simpler.**

a + 6 is called an expression.
b = *a* + 6 is called a formula. The value of *b* depends on the value of *a*.

Collecting like terms

Expressions can be simplified by collecting like terms.

You can only collect together terms that include exactly the same letter combinations.

Examples

1. $5a + 3a = 8a$

2. $3a - 4b + 2a + 3b = 5a - b$

    Do $3a + 2a$ first and then $-4b + 3b$, which is $-1b$ or $-b$.

3. $5a^2 + 3a^2 - 2a^2 = 6a^2$

4. $5a - 3b$ cannot be simplified

5. $2ab + 3ba = 5ab$

Writing formulae

Example

My brother is 3 years older than me. My mother is 3 times as old as me.

If I am *n* years old, write expressions for my brother's and mother's ages.

If I am 15, my brother is $15 + 3 = 18$

So if I am *n* years old my brother is $n + 3$ years old.

If I am 15, my mother is $3 \times 15 = 45$

So if I am *n* years old my mother is $3 \times n$ or $3n$ years old.

Write down a formula for the sum (*s*) of the ages of me, my mother and brother.

$s = n + n + 3 + 3n$

$s = 5n + 3$

FORMULAE AND EXPRESSIONS 2

○ Substituting into formulae

Replacing a letter with a number is called **substitution**.

> ● Write out the expression first and then replace the letters with the values given.
>
> ● Work out the value – but take care with the order of operations, i.e. BIDMAS.

Examples

1. $a = 3b - 4c$. Find a if $b = 4$ and $c = -2$.

   $a = (3 \times 4) - (4 \times -2)$

   $= 12 - (-8)$ Taking away a negative is the same as adding.

   $= 20$

2. $E = \frac{1}{2} mv^2$. Find E if $m = 6$ and $v = 10$.

   $= \frac{1}{2} \times 6 \times 10^2$

   $= \frac{1}{2} \times 6 \times 100$

   $E = 300$

3. $V = u + at$. Find V if $u = 22$, $a = -2$ and $t = 6$.

   $= 22 + (-2 \times 6)$

   $= 22 + (-12)$

   $= 22 - 12$

   $= 10$

Rearranging formulae

The subject of a formula is the letter that appears on its own on one side of the formula.

Examples

1. Make a the subject of the formula $b = (a - 3)^2$.

$b = (a - 3)^2$	Deal with the power first. Square root both sides.
$\sqrt{b} = a - 3$	Remove any term added or subtracted. Add 3 to both sides.

   $\sqrt{b} + 3 = a$

   $a = \sqrt{b} + 3$

2. Make x the subject of the formula $5(y + x) = 8x + 3$

   When the subject occurs on both sides of the equal sign, they need to be collected on one side.

   $5(y + x) = 8x + 3$

   $5y + 5x = 8x + 3$

   $5y - 3 = 8x - 5x$

   $5y - 3 = 3x$

   $x = \dfrac{5y - 3}{3}$

Progress check

1. Simplify the following expressions:

   a) $6a - 3b + 2a - 4b$

   b) $3a^2 - 6b - 2b + a^2$

   c) $5xy - 3yx + 2xy^2$

2. If $a = \dfrac{3}{5}$ and $b = -2$, find the value of these expressions, giving your answer to 3sf where appropriate.

   a) $ab - 5$

   b) $a^2 + b^2$

   c) $3a - 6ab$

3. Make u the subject of the formula $v^2 = u^2 + 2as$.

Single brackets

> Each term outside the brackets multiplies each separate term inside the bracket.

Examples

Expand and simplify:

a) $5(x + 6) = 5x + 30$

b) $-2(2x + 4) = -4x - 8$

c) $3(2x - 5) - 2(x - 3)$

   $= 6x - 15 - 2x + 6$ Multiply out the brackets.

   $= 4x - 9$ Collect like terms.

Two brackets

$(x + 4)(x + 2)$ $= x^2 + 2x + 4x + 8$

   $= x^2 + 6x + 8$

- Every term in the second bracket must be multiplied by every term in the first bracket.

- Often, but not always, the two middle terms are like terms and can be collected together.

Examples

$(x + 4)(2x - 5) = 2x^2 - 5x + 8x - 20$

   $= 2x^2 + 3x - 20$

$(2x + 1)^2 = (2x + 1)(2x + 1)$

   $= 4x^2 + 2x + 2x + 1$

   $= 4x^2 + 4x + 1$ Remember that x^2 means x times by itself.

Multiplying out brackets helps to simplify algebraic expressions.

○ Factorisation

> Factorisation simply means putting into brackets.

One bracket

$4x + 6 = 2(2x + 3)$

To factorise $4x + 6$

- Recognise that 2 is the highest common factor of 4 and 6.

- Take out the common factor.

- The expression is completed inside the bracket so that when multiplied out it is equivalent to $4x + 6$.

Two brackets

Two brackets are obtained when a quadratic expression of the type $ax^2 + bx + c$ is factorised.

Examples

$x^2 + 4x + 3 = (x + 1)(x + 3)$
$x^2 - 7x + 12 = (x - 3)(x - 4)$
$x^2 + 3x - 10 = (x + 5)(x - 2)$

Progress check

1 Expand and simplify:
   a) $(x + 3)(x - 2)$
   b) $2(3x - 4)$
   c) $4x(x - 3)$
   d) $(x - 3)^2$

2 Factorise:
   a) $4x^2 + 8x$
   b) $12xy - 6x^2$
   c) $3a^2b + 6ab^2$

3 Factorise:
   a) $x^2 + 4x + 4$
   b) $x^2 - 5x + 6$
   c) $x^2 - 4x - 5$

SOLVING LINEAR EQUATIONS

○ Type 1 of the form $ax + b = c$

Example

Solve:

$5x - 2 = 13$

$5x = 13 + 2$ Add 2 to both sides.

$5x = 15$

$x = 15 \div 5$ Divide both sides by 5.

$x = 3$

> Remember to add the same thing to both sides of the equation so that they balance.

○ Type 2 of the form $ax + b = cx + d$

Example

Solve: $7x - 4 = 3x + 8$

$7x = 3x + 12$ Add 4 to each side.

$4x = 12$ Subtract $3x$ from both sides.

$x = 12 \div 4$

$x = 3$

Must check:

$7 \times 3 - 4 = 3 \times 3 + 8$

$21 - 4 = 9 + 8$ ✔

Yes, it's right.

○ Type 3 with brackets!

Examples

i)   Solve:

$5(x - 1) = 3(x + 2)$

$5x - 5 = 3x + 6$

$5x = 3x + 11$

$2x = 11$

$x = 5.5$

Just multiply out the brackets and solve as normal.

ii)   Solve:

$\dfrac{3(2x - 1)}{5} = 6$

$3(2x - 1) = 6 \times 5$

$6x - 3 = 30$

$6x = 33$

$x = 33 \div 6$

$x = 5.5$

Multiply both sides by 5.

○ Type 4 problems

Always write down the information that you know.

The perimeter of this rectangle is 30 cm. Work out the value of y and find the length of the rectangle.

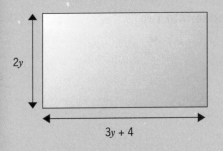

2y

3y + 4

Write down what you know.

$3y + 4 + 2y + 3y + 4 + 2y = 30$

Simplify the expression and solve as normal.

$10y + 8 = 30$

$\quad 10y = 30 - 8$

$\quad 10y = 22$

$\quad\quad y = 2.2$

Length of rectangle $= 3 \times 2.2 + 4$

$\quad\quad\quad\quad\quad\quad = 10.6$ cm

Progress check

Solve the following equations:

1 $2x - 6 = 10$

2 $5 - 3x = 20$

3 $4(2 - 2x) = 12$

4 $6x + 3 = 2x - 10$

5 $7x - 4 = 3x - 6$

6 $5(x + 1) = 3(2x - 4)$

7 The perimeter of this triangle is 60 cm. Work out the value of x and find the shortest length.

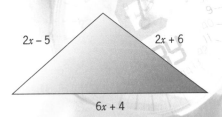

2x – 5   2x + 6

6x + 4

● Solving quadratic equations

Solve: $x^2 - x - 6 = 0$

Must check that the equation equals zero.

It's a quadratic of the form $ax^2 + bx + c = 0$

Need to factorise into two brackets ( )( ) = 0

DAY
3

$$x^2 - x - 6 = 0$$

$$(x + 2)(x - 3) = 0$$ Factorise into 2 brackets.

either $(x + 2) = 0$ Since the equation equals zero one of the
or $(x - 3) = 0$ brackets must equal zero.

hence $x = -2$ or $x = 3$

Example

Solve: $x^2 + 5x + 6 = 0$

          $(x + 2)(x + 3) = 0$

either $(x + 2) = 0$

or $(x + 3) = 0$

so $x = -2$ or $x = -3$

Solving cubic equations

Trial and improvement gives an approximate solution to cubic equations.

Example

The equation $x^3 + 2x = 58$ has a solution between 3 and 4. Find the solution to 1 decimal place.

Drawing a table can help you – but also the examiner – since it makes it easier to follow what you have done.

x	$x^3 + 2x$	Comment
3.5	$3.5^3 + 2 \times 3.5 = 49.875$	too small
3.8	$3.8^3 + 2 \times 3.8 = 62.472$	too big
3.7	$3.7^3 + 2 \times 3.7 = 58.053$	too big
3.65	$3.65^3 + 2 \times 3.65 = 55.927$	too small
		$x = 3.7$ (1dp)

Progress check

1. Decide whether the answers given are correct.

   Solve fully:

   a) $x^2 + 4x + 3 = 0$,

   $x = -3, x = -1$

   b) $x^2 + 5x = 0$, $x = -5$

   c) $x^2 - 2x + 1 = 0$, $x = 1$

   d) $x^2 - x - 20 = 0$,

   $x = -4, x = 5$

2. Solve:

   a) $x^2 - 7x = 0$

   b) $x^2 + 8x + 15 = 0$

   c) $x^2 - 5x + 6 = 0$

3. The equation $a^3 = 40 - a$ has a solution between 3 and 4.
   Find the solution to 1dp by using a method of trial and improvement.

DAY
3

SIMULTANEOUS EQUATIONS

By algebra (elimination method)

Solve simultaneously:	$3x + 2y = 8$
	$2x - 3y = 14$

Label the equations ① and ②.	$3x + 2y = 8$ ①
	$2x - 3y = 14$ ②

Since no coefficients match, multiply equation ① by 2 and equation ② by 3.	$6x + 4y = 16$
	$6x - 9y = 42$

Rename them equations ③ and ④.	$6x + 4y = 16$ ③
	$6x - 9y = 42$ ④

The coefficient of x in equations ③ and ④ is the same. Subtract equation ④ from equation ③ and solve remaining equation.	$0x + 13y = -26$
	$y = -26/13$
	$y = -2$

Substitute the value of $y = -2$ back into equation ①. Solve this equation to find x.	$3x + (-4) = 8$
	$3x = 8 + 4$
	$3x = 12$
	$x = 4$

Check in equation ②.	$(2 \times 4) - (3 \times -2) = 14$ ✔

Solution is: $x = 4$, $y = -2$

Graphically

Solve the simultaneous equations

$2x + 3y = 6$
$x + y = 1$

> **The point at which any two graphs intersect represents the simultaneous solutions of their equations.**

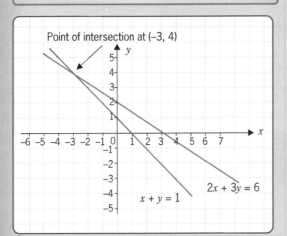

Point of intersection at (−3, 4)

$2x + 3y = 6$
$x + y = 1$

Draw the graph of:

$2x + 3y = 6$
$x = 0 \ 3y = 6 \therefore y = 2 \ (0, 2)$
$y = 0 \ 2x = 6 \therefore x = 3 \ (3, 0)$

Draw the graph of:

$x + y = 1$
$x = 0, y = 1 \therefore (0, 1)$
$y = 0, x = 1 \therefore (1, 0)$

At the point of intersection: $x = −3, y = 4$

Progress check

1 Solve the following pairs of simultaneous equations:

a) $4b + 7a = 10$
$2b + 3a = 3$

b) $2p + 3r = 6$
$p + r = 1$

c) $y − 2x = −1$
$x + y = 5$

2 The diagram shows the graphs of the equations:

$x + y = 6$ and $y = x + 2$

Use the diagram to solve the simultaneous equations

$x + y = 6$
$y = x + 2$

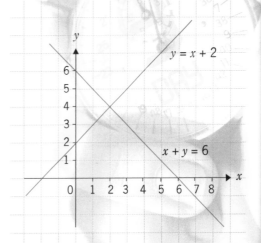

$y = x + 2$

$x + y = 6$

DAY
3

10 MINS

DAY 3

Special sequences

Odd numbers
1, 3, 5, 7, 9 ...
nth term is $2n - 1$

Even numbers
2, 4, 6, 8, 10 ...
nth term is $2n$

Square numbers

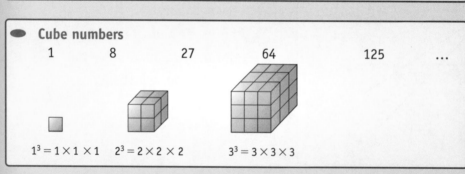

| 1 | 4 | 9 | 16 | 25 | ... |

$1^2 = 1 \times 1$ $2^2 = 2 \times 2$ $3^2 = 3 \times 3$ $4^2 = 4 \times 4$ $5^2 = 5 \times 5$

Cube numbers

| 1 | 8 | 27 | 64 | 125 | ... |

$1^3 = 1 \times 1 \times 1$ $2^3 = 2 \times 2 \times 2$ $3^3 = 3 \times 3 \times 3$

Triangle numbers

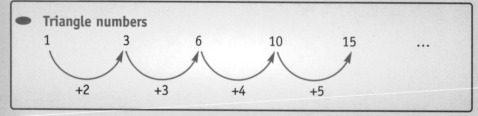

1   3   6   10   15   ...

+2   +3   +4   +5

Fibonacci sequence
1, 1, 2, 3, 5, 8, 13 ...   Add the previous two terms.

SPEND 10 MINUTES ON THIS TOPIC

Finding the *n*th term of a linear sequence

The *n*th term is often denoted by **U*n***. For example the 8th term is U_8.

For a linear sequence the nth term takes the form:

$$U_n = an + b$$

Example

Find the *n*th term of this sequence:

2,   6,   10,   14

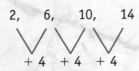

+ 4   + 4   + 4

- See how the numbers are jumping (going up in 4s).
- The *n*th term is 4*n* + or − something.
- Try out 4*n* on the first term. This gives $4 \times 1 = 4$, but the first term is 2 ... so subtract 2.
- The rule is 4*n* − 2.
- Test this rule on the other terms
  $1 \rightarrow 4 - 2 = 2$
  $2 \rightarrow 8 - 2 = 6$
  $3 \rightarrow 12 - 2 = 10$
  It works on all of them.
- *n*th term is 4*n* − 2
- The 20th term in the sequence would be: $4 \times 20 - 2 = 78$

Progress check

1 The cards show the *n*th term of some sequences:

| 2*n* | 4*n* + 1 | 3*n* + 2 | 5*n* − | 2 − *n* |

Match the cards with the sequences below:

a)   5, 9, 13, 17, ...
b)   1, 0, −1, −2, ...
c)   2, 4, 6, 8, 10, ...
d)   5, 8, 11, 14, 17, ...
e)   4, 9, 14, 19, ...

2 Find the *n*th term of these sequences:

a)   7, 10, 13, 16, 19, ...
b) $\frac{1}{3}, \frac{1}{5}, \frac{1}{7}, \frac{1}{9}, ...$
c)   1, 4, 7, 10 ...

INEQUALITIES

Inequalities can be solved in exactly the same way as equations except that when multiplying or dividing by a negative number, you must reverse the inequality sign.

● The inequality symbols

> means **greater than**

≥ means **greater than or equal to**

< means **less than**

≤ means **less than or equal to**

Example

Solve:

$2x - 2 < 10$

$2x < 10 + 2$

$2x < 12$

$x < 6$

Solve:

$3 - 2x \geqslant 9$

$-2x \geqslant 9 - 3$

$-2x \geqslant 6$

$x \leqslant \dfrac{6}{-2}$

$x \leqslant -3$ Divide by −2 and change inequality sign round.

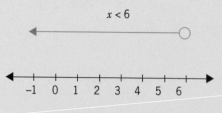

The open circle means that 6 is not included.

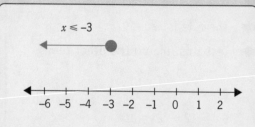

The solid circle means that −3 is included.

DAY 3

Example

Solve:

$-2 < 4x - 3 \leqslant 9$ The integer values
$1 < 4x \leqslant 12$ that satisfy this
$\frac{1}{4} < x \leqslant 3$ inequality are 1, 2, 3.

● Graphs of inequalities

The graph of an equation such as $x = 2$ is a line, whereas the graph of the inequality $x < 2$ is a region that has $x = 2$ as its boundary.

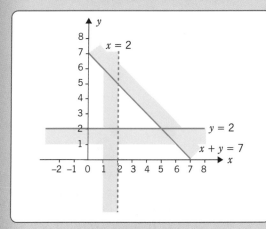

The diagram shows unshaded the region:

$x + y \leqslant 7$ For strict inequalities $>$
$x > 2$ and $<$ the boundary line
$y \geqslant 2$ is not included and is
  shown as a dashed line.

Progress check

1 Solve the following inequalities:

  a) $5x - 1 < 10$

  b) $6 \leq 3x + 2 < 11$

  c) $3 - 5x < 12$

2 On the diagram below, leave unshaded the region satisfied by these inequalities:

$x + y \leqslant 5$

$x \geqslant 1$

$y > 1$

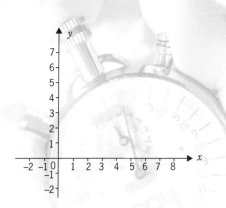

DAY

3

6 Label the graph once you've drawn it.

x	−1	0	2	4
y	−7	−4	2	8

1 To work out the coordinates of the points that lie on the line $y = 3x - 4$, draw a table of values.

5 The graph $y = 3x - 4$ is drawn.

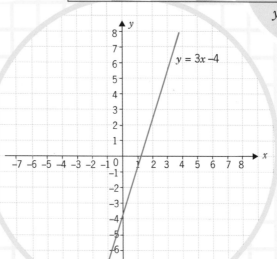

$y = 3x - 4$

4 Join the points with a straight line.

3 The coordinates of the points on the line are:
(−1, −7) (0, 4)
(2, 2) (4,8)
Just read them from the table of values.

2 Substitute the x values into the equation $y = 3x - 4$, to find the values of y i.e. $x = 2$,
$y = 3 \times 2 - 4 = 2$

The general equation of a straight lined graph is
$$y = mx + c$$
m is the gradient and c is the intercept on the y-axis.

Gradient of a straight line

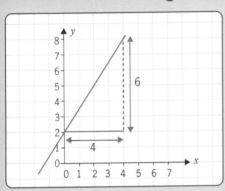

$$\text{Gradient} = \frac{\text{change in } y}{\text{change in } x} \quad \text{OR} \quad \frac{\text{height}}{\text{base}}$$

$$\text{Gradient} = \frac{6}{4} = \frac{3}{2} \text{ or } 1.5$$

Be careful when finding the gradient – double-check the scales.

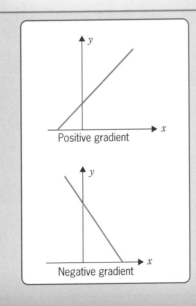

Positive gradient

Negative gradient

How are you feeling Grandad?

Look at my graph – I'm getting worse each day.

I know how to make you feel a bit better.

Your graph has slipped – you're fine !

DAY 3

CURVED GRAPHS

Draw the graph of $y = x^2 - 2x - 6$. Use values of x from -2 to 3.

STEP 1 Draw a table of values.

x	-2	-1	0	1	2	3
y	2	-3	-6	-7	-6	-3

STEP 2 Fill in the table of values by substituting the values of x into the equation.

i.e. $x = 1$ $y = 1^2 - 2 \times 1 - 6, y = -7$

Coordinates are $(1, -7)$

STEP 3 Draw the axes on graph paper and plot the points.

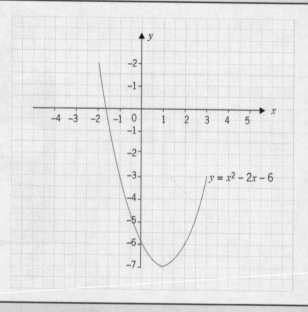

$y = x^2 - 2x - 6$

STEP 4 Join the points with a smooth curve.

STEP 5 Label the curve.

SPEND 15 MINUTES ON THIS TOPIC

The minimum value is when $x = 1$, $y = -7$

The line of symmetry is at $x = 1$

Other graph shapes

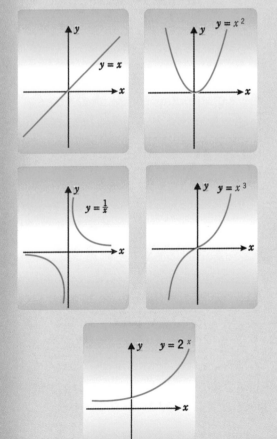

Progress check

1 a) Complete the table of values for $y = x^3 - 1$.

x	−3	−2	−1	0	1	2	3
y							

b) Draw the graph of $y = x^3 - 1$. Use scales of 2 units for 1 cm on the x-axis and 20 units for 1 cm on the y-axis.

c) From the graph find the value of x when $y = 15$.

2 Match each graph below to one of the equations.

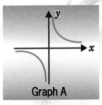

Graph A

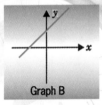

Graph B

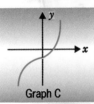

Graph C

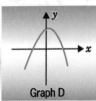

Graph D

$y = x^3 - 5$

$y = 2 - x^2$

$y = 4x + 2$

$y = \dfrac{3}{x}$

DAY 4

Mr Smith travels from St Albans to his office 80 miles away.

The travel graph shows his journey.

Part B shows that Mr Smith was stationary for 1 hour between 1100 and 1200.

Part E shows the return journey at a constant speed of 40 mph for 2 hours.

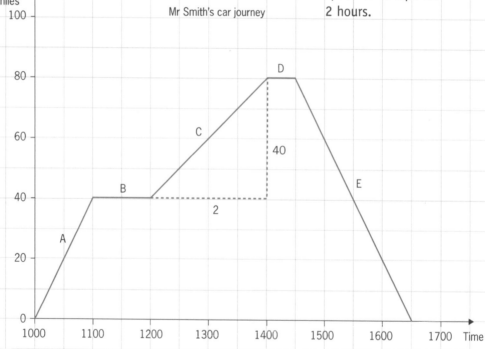

Mr Smith's car journey

Part A shows Mr Smith travels at a constant speed of 40 mph.

Part C shows that Mr Smith travels at a constant speed of 20 mph

Part D shows that Mr Smith was stationary for 30 minutes.

Key points

- Always check that you understand the scales before starting. For this graph

  - vertically 1 square represents 10 miles

  - horizontally 2 squares represents 1 hour

- The gradient of the distance–time graph represents the speed over that time interval.

$$\text{speed} = \frac{\text{distance travelled}}{\text{time taken}}$$

Example: In part C of the graph

$\text{speed} = \frac{40}{2}$

$\text{speed} = 20\,\text{mph}$

Progress check

The travel graph shows the car journeys of two people.

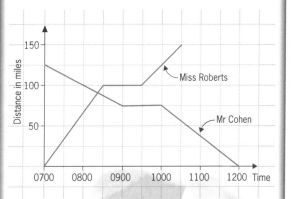

Decide whether these statements are true or false.

1 Miss Roberts travelled the first 100 miles at a speed of 50 mph.

2 Mr Cohen had a rest for 1 hour between 0900 and 1000.

3 Mr Cohen travelled at a speed of 37.5 mph between 1000 and 1200.

4 Miss Roberts and Mr Cohen pass each other at 0820.

5 Miss Roberts travelled at a speed of 50 mph between 0930 and 1030.

Constructing a triangle

Use compasses to construct this triangle.

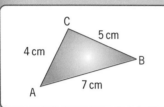 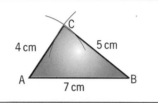

- Draw the longest side AB.

- With the compass point at A, draw an arc of radius 4 cm.

- With the compass point at B, draw an arc of radius 5 cm.

- Join A and B to the point where the two arcs meet at C.

The perpendicular bisector of a line

- Draw a line XY.

- Draw two arcs with the compasses, using X as the centre. The compasses must be set at a radius greater than half the distance of XY.

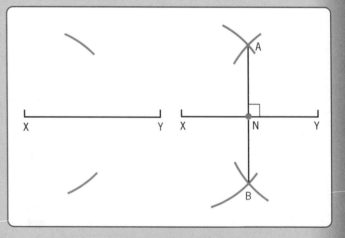

- Draw two more arcs with Y as the centre.

  (Keep the compasses the same distance apart as before.)

- Join the two points where the arcs cross.

- AB is the **perpendicular bisector** of XY.

- N is the **midpoint** of XY.

The following constructions can be completed using only a ruler and a pair of compasses.

The perpendicular from a point to a line

- From P draw arcs to cut the line at A and B.

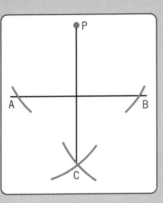

- From A and B draw arcs with the same radius to intersect at C.

- Join P to C; this line is perpendicular to AB.

Bisecting an angle

- Draw two lines XY and YZ to meet at an angle.

- Using compasses, place the point at Y and draw arcs on XY and YZ.

- Place the compass point at the two arcs on XY and YZ and draw arcs to cross at N. Join Y and N. YN is the **bisector** of angle XYZ.

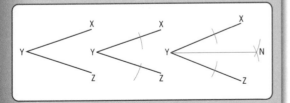

Progress check

1 Construct an angle of 30°, using ruler and pair of compasses only.

2 Draw the perpendicular bisector of an 8 cm line.

3 Bisect this angle.

LOCI

Types of loci

1. The locus of the points that are a constant distance from a fixed point is a circle.

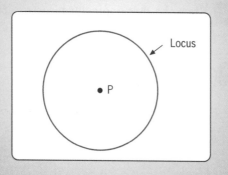

2. The locus of the points that are equidistant from two points XY is the perpendicular bisector of XY.

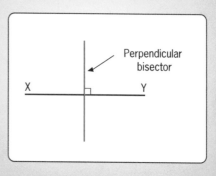

3. The locus of the points that are equidistant from two lines is the line that bisects the angle between the lines.

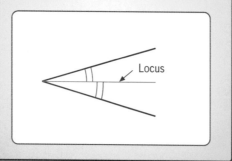

4. The locus of the points that are a constant distance from a line XY is a pair of parallel lines above and below XY.

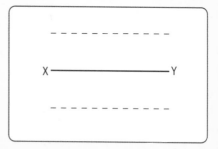

Sometimes you need to combine types 1 and 3.

A fixed distance from a line segment gives this locus.

Example

Three radio transmitters form an equilateral triangle ABC with sides of 50 km. The range of the transmitter at A is 37.5 km, at B 30 km and at C 28 km. Using a scale of 1 cm to 10 km, construct a scale diagram to show where signals from all three transmitters can be received.

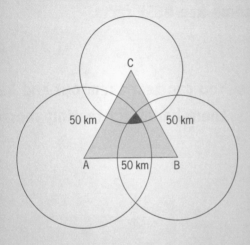

Please note that on your scale drawing the circle at A would have a radius of 3.75 cm. The circle at B would have a radius of 3 cm and the circle at C a radius of 2.8 cm.

Progress check

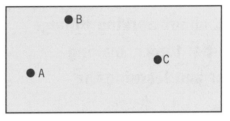

The plan shows a garden drawn to a scale of 1 cm: 2 m. A and B are bushes and C is a pond. A landscape gardener has decided:

a) to lay a path right across the garden at an equal distance from each of the bushes.

b) to lay a flower border 4 m wide around pond C.

Construct these features on the plan above.

HOW TO USE THE QUIZ CARDS

There are several stages to successful revision – one of the most important is writing a list of the topics you need to know.

Then it's all about working through these essential topics, making useful notes and learning the key facts.

This is where these quiz cards can help you.

The questions on the cards provide a last-minute check of some key GCSE facts.

- You can leave them in the book and refer to them when you want

- You can tear them out and keep them handy for testing yourself

- You can get someone else to test you

- You can test your friends, which is also a good way of helping information sink in

- You can add to the cards by making your own sets of questions and answers

Remember – PREPARATION and PRACTICE and you'll be on the way to a good result!

What is the formula for the area of a circle?

What is 0.27 written in standard form?

What are the three things you need to remember about bearings?

Simplify $a^4 \times 3a^7$

What is 10% of £425?

What is the name of the quadrilateral with one pair of parallel sides?

The probability that it rains tomorrow is 0.3. What is the probability that it does not rain tomorrow?

Work out the answer to 0.3×0.4.

What is 7 out of 25 as a percentage?

A triangle has lengths 3 cm, 4 cm and 5 cm. Is it right-angled?

2.7×10^{-1}	πr^2
$3a^{11}$	measured from the North clockwise direction three figures
Trapezium	£42.50
0.12	0.7
Yes since $3^2 + 4^2 = 5^2$	28%

The weight of an object is 42 kg, to the nearest kilogram. What is the minimum weight the object could be?

If the equation of a graph is $2y = 4x + 8$, what is the gradient of the line?

Multiply out these brackets: $3x(x + 2)$

What is 4^0 equal to?

The probability of getting a "Bulls eye" in darts is 0.12. Out of 200 throws, how many would you expect to hit the 'Bulls eye?'

Add together $\frac{2}{7} + \frac{4}{5}$.

If $v = u + at$, what is the value of v if $u = 4$, $a = 2$ and $t = -3$?

How many pounds are in 4 kilograms?

4 kg

Factorise $5xy^2 + 10x^2y$

If a and b are variables, what does the formula $2a^2b + 3ab^2$ represent, length, area or volume?

gradient $= 2$	41.5 kg
1	$3x^2 + 6x$
$1\frac{3}{35}$ or $\frac{38}{35}$	24
8.8 pounds	-2
Volume	$5xy(y + 2x)$

EXAM TECHNIQUE

FOLLOW OUR CHECKLIST TO HELP YOU BEFORE AND DURING THE EXAMS

Preparation

Use the time before the exams effectively. Write a list of all the topics you have to cover. Work through your notes systematically and ask for help with any topics that you're struggling to understand.

Practice

Attempt as many practice questions and past papers as possible. Familiarise yourself with the question types, the marks allocated and the time allowed. Compare your marks to those given in the mark schemes – see where you did well and where there is room for improvement.

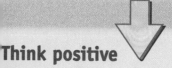

Think positive

Even if time is running short, remind yourself of the progress you have made. Use what time is left by working through the key topics – either those that are most likely to come up in the exam or those that you find most difficult.

IN THE EXAM ITSELF...

- Follow all the instructions in the exam paper
- Attempt the correct number of questions
- Read each question carefully and more than once

- Highlight the key words in the question and note the command word – State, Describe, Explain, Discuss, Find, Suggest, Calculate, List etc.
- Check the number of marks available for each question and answer accordingly

- Plan your response in brief note form
- Ensure that you answer the question asked and that your response stays relevant
- Allocate time carefully and make sure you complete the paper

- Return to any questions you have left out and read through your answers at the end
- Remember that accurate spelling and good use of English do count

We hope this book will help you on the way to GCSE success.

ANGLES

Angles on a straight line add up to 180°.
$a + b + c = 180°$

An exterior angle of a triangle equals the sum of the two opposite interior angles.
$a + b = c$

Vertically opposite angles are equal.
$a = b, c = d$
$a + d = b + c = 180°$

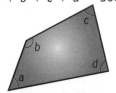

ANGLE FACTS

Angles at a point add up to 360°.
$a + b + c + d = 360°$

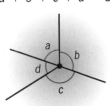

Angles in a triangle add up to 180°.
$a + b + c = 180°$

Angles in quadrilateral add up to 360°.
$a + b + c + d = 360°$.

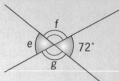

Examples

Find the missing angles in the diagrams below:

1.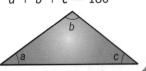

2.

$a = 64°$ (opposite angles)
$b = 180° - 64°$ (angles on a straight line)
$b = 116°$
$c = 64°$ (isosceles triangle)
$d = 52°$ (angles in a triangle)

$e = 72°$ (opposite angles)
$f = 108°$ (angles on a straight line)
$g = 108°$

Whenever lines meet or intersect, the angles they make follow certain rules.

15 MINS

Parallel lines

Three types of relationships are produced when a line called a transversal crosses a pair of parallel lines.

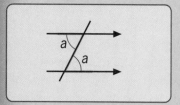

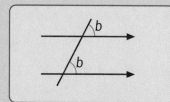

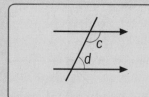

Alternate (z) angles are equal.

Corresponding angles are equal.

Supplementary angles add up to 180°. $c + d = 180°$

Angles in polygons

There are two types of angles in a polygon – **interior** or **exterior**.

For a regular polygon with n sides:

- Size of an exterior angle = $\frac{360°}{n}$
- Interior angle + exterior angle = $180°$
- Sum of interior angles is $(n - 2) \times 180°$ or $(2n - 4) \times 90°$

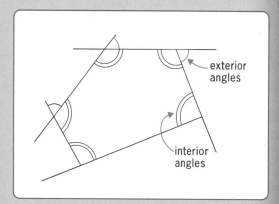

exterior angles

interior angles

Example

A regular polygon has an interior angle of 108°. How many sides does it have?

$180° - 108° = 72°$ (size of exterior angle)
$360° \div 72° = 5$ (number of sides)

DAY 4

Compass directions

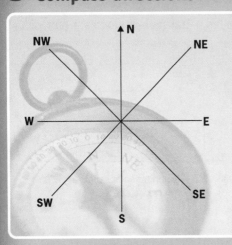

Bearings

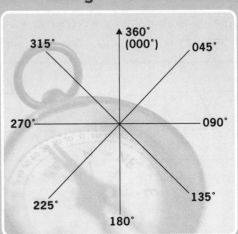

Examples

Find the bearings of P **from** Q in each diagram below:

a)

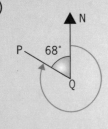

b)

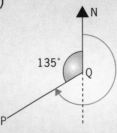

c)

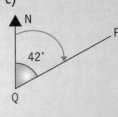

Bearing of P from Q

$= 360° - 68°$

$= 292°$

Bearing of P from Q

$= 360° - 135°$

$= 225°$

Bearing of P from Q

$= 042°$

Back bearings are more difficult – you need to draw in a second North line. The angle properties of parallel lines can then be used.

There are 3 important key facts about bearings:
- they are always measured from the North (N)
- they are measured in a clockwise direction
- they are written using 3 figures

Examples

Find the bearings of Q **from** P in each diagram below:

a)

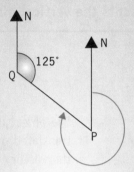

b)

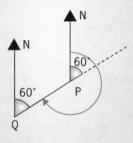

a) Bearing of Q from P

$= 360° − 55°$

$= 305°$

b) Bearing of Q from P

$= 60° + 180°$

$= 240°$

Bearings are often used in scale drawing questions.

Progress check

1 Fill in the statements with the correct words:

a) A bearing is measured from the .... in a .... direction.

b) A bearing is written using .... figures.

c) The bearing of the compass direction .... is 315 degrees.

2 For the following diagrams find the bearing of B from A.

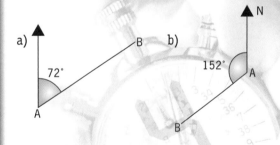

a)

72°

b)

152°

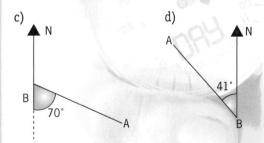

c)

N

B

70°

A

d)

A

41°

N

B

DAY
4

Translations

Translations move figures from one position to another position. **Vectors** are used to describe the distance and direction of the translations.

A vector is written as $\begin{pmatrix} a \\ b \end{pmatrix}$

a represents the horizontal distance

b represents the vertical distance

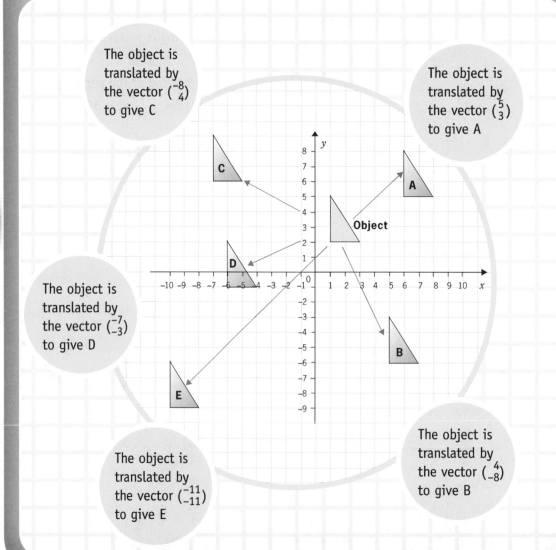

The object is translated by the vector $\begin{pmatrix} -8 \\ 4 \end{pmatrix}$ to give C

The object is translated by the vector $\begin{pmatrix} 5 \\ 3 \end{pmatrix}$ to give A

The object is translated by the vector $\begin{pmatrix} -7 \\ -3 \end{pmatrix}$ to give D

The object is translated by the vector $\begin{pmatrix} -11 \\ -11 \end{pmatrix}$ to give E

The object is translated by the vector $\begin{pmatrix} 4 \\ -8 \end{pmatrix}$ to give B

There are four types of transformations: reflection, rotation, translation and enlargement.

> The object and the image are congruent when the shape is translated.

⚬ Reflections

These create an image of an object on the other side of the mirror line.

The mirror line is known as an **axis of reflection**.

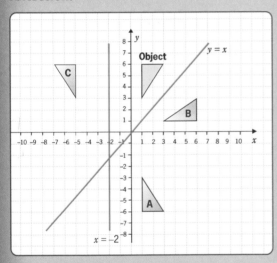

The object is reflected in the x-axis (or $y = 0$) to give the image A.

The object is reflected in the line $y = x$ to give the image B.

The object is reflected in the line $x = -2$ to give the image C.

a little bit of reflection...

mirror, mirror on the wall...

Rotations

In a rotation the object is turned by a given angle about a fixed point called the **centre of rotation**. The size and shape of the figure are not changed.

The object A is rotated by 90° clockwise about (0, 0) to give the image B.

The object A is rotated by 180° about (0, 0) to give the image C.

The object A is rotated 90° anticlockwise about (–2, 2) to give the image D.

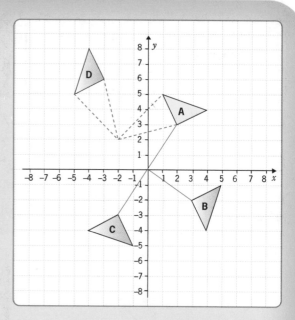

Enlargements

These change the size but not the shape of the object. The **centre of enlargement** is the point from which the enlargement takes place. The **scale factor** indicates how many times the lengths of the original figure have changed size.

Example

This shape has been enlarged by a scale factor of 4.
Centre of enlargement at (0, 0).

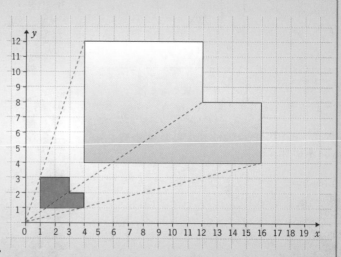

Each side of the enlargement is 4 times the size of the original.

After a rotation, the image is congruent to the object. After an enlargement, the enlarged shape is similar to the object.

Example

Describe fully the transformation that maps ABC onto A'B'C'.

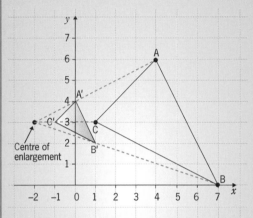

- To find the centre of enlargement join A to A', B to B' etc. and continue the line.

- Where all the lines meet is the centre of enlargement: (−1, 3)

- The transformation is an enlargement of scale factor $\frac{1}{3}$. Centre of enlargement is (−2, 3).

An enlargement with a scale factor less than 1 makes the shape smaller.

Progress check

For the diagram below, describe fully the transformation that maps:

1. A onto B

2. B onto C

3. A onto C

4. A onto D

5. A onto E

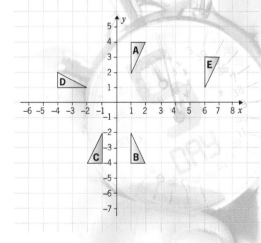

$$a^2 + b^2 = c^2$$

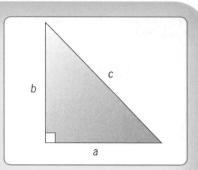

○ Finding the hypotenuse (n)

$n^2 = 7^2 + 12^2$	square the two sides
$n^2 = 49 + 144$	add the two sides together
$n^2 = 193$	
$n = \sqrt{193}$	square root
$n = 13.9\,\text{cm}$ (3sf)	round to 3sf

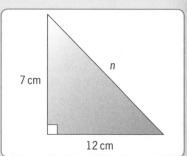

○ Finding a short side (p)

$15^2 = p^2 + 8^2$	
$15^2 - 8^2 = p^2$	when finding a shorter length remember to subtract
$225 - 64 = p^2$	
$161 = p^2$	
$\sqrt{161} = p$	
$p = 12.7\,\text{cm}$ (3 sf)	

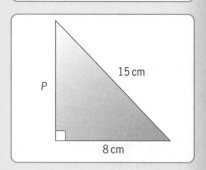

○ Finding the length of a line AB, given the coordinates of its end points

Horizontal distance $= 6 - 1 = 5$

Vertical distance $= 5 - 2 = 3$

Length of $(AB)^2 = 5^2 + 3^2$

$= 25 + 9$

$= 34$

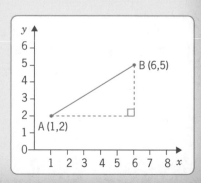

The hypotenuse is the longest side of a right-angled triangle. It is always opposite the right angle.

15 MINS

$AB = \sqrt{34}$

$AB = 5.83\,\text{cm}$

> We could leave this as $\sqrt{34}$. This is known as leaving in **surd form**.

Solving a more difficult problem

Calculate the vertical height of this isosceles triangle.

8 cm   8 cm

11 cm

Remember to split the triangle down the middle to make it right-angled.

Using Pythagoras' theorem gives:

$$8^2 = h^2 + 5.5^2$$

$$64 = h^2 + 30.25$$

$$64 - 30.25 = h^2$$

$$33.75 = h^2$$

$$\sqrt{33.75} = h^2$$

$$h = 5.81\,\text{cm (3sf)}$$

8 cm   h

5.5 cm

Progress check

1 Molly says: 'The angle x in this triangle is 90°.'
Explain how Molly knows that without measuring the size of the angle.

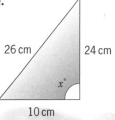

26 cm   24 cm

$x°$

10 cm

2 Calculate the length of the diagonal of this rectangle. Give your answer to one decimal place (1dp).

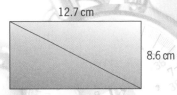

12.7 cm

8.6 cm

3 Calculate the length of CD in this diagram.

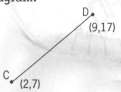

D
(9,17)

C
(2,7)

Leave your answer in surd form.

In similar shapes:
- Corresponding angles are equal.
- Corresponding lengths are in the same ratio.

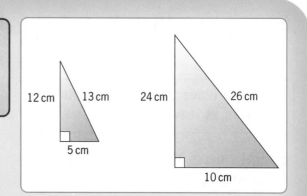

Finding missing lengths of similar figures

These questions are very common at GCSE.

Examples

Find the missing length labelled a in the diagrams below:

a)

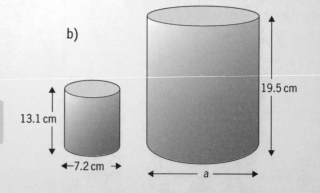

$\dfrac{a}{12} = \dfrac{3.8}{8.5}$ Corresponding lengths are in the same ratio.

$a = \dfrac{3.8}{8.5} \times 12$ Multiply both sides by 12.

$a = 5.36$ cm (3sf)

b)

$\dfrac{a}{7.2} = \dfrac{19.5}{13.1}$

$a = \dfrac{19.5}{13.1} \times 7.2$ Multiply both sides by 7.2.

$a = 10.7$ cm (3sf)

Objects that are exactly the same shape but different sizes are called similar shapes. One is an enlargement of the other.

15 MINS

Example

Calculate the missing length y.

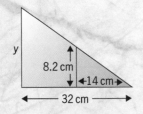

- Firstly draw out the individual triangles:

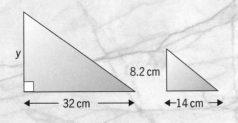

- Write down the corresponding ratios:

$$\frac{y}{32} = \frac{8.2}{14}$$

- Multiply both sides by 32:

$$y = \frac{8.2}{14} \times 32$$

$$= 18.7 \text{ cm}$$

This gives an alternative way to writing the ratios as seen in the other examples. Both are correct!

Progress check

1. Are these two triangles similar?

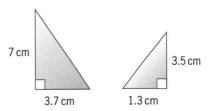

2. Calculate the lengths marked n in these similar shapes. Give your answers correct to 1dp.

a)

b)

c)

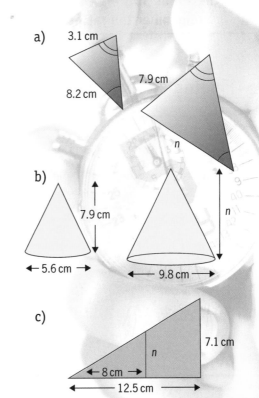

DAY 5

The sides of a right-angled triangle are given temporary names according to where they are in relation to a chosen angle θ.

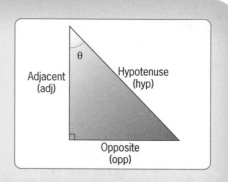

The trionometric ratios are:

$$\text{Sine } \theta = \frac{\text{Opposite}}{\text{Hypotenuse}} \quad \text{Cosine } \theta = \frac{\text{Adjacent}}{\text{Hypotenuse}} \quad \text{Tangent } \theta = \frac{\text{Opposite}}{\text{Adjacent}}$$

Use the words **SOH – CAH – TOA** to remember the ratios.

Example: TOA means $\tan \theta = \dfrac{\text{opp}}{\text{adj}}$

Finding a length

Example

Find the missing length y in the diagram.

- Label the sides first.

- Decide on the ratio.

  $\sin 30° = \dfrac{\text{opp}}{\text{hyp}}$

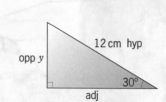

- Substitute in the values

  $\sin 30° = \dfrac{y}{12}$

  $12 \times \sin 30° = y$ Multiply both sides by 12.

  $y = 6\,\text{cm}$

Trigonometry in right-angled triangles can be used to find an unknown angle or length.

⊂ Finding an angle

Example

Calculate angle $A\hat{B}C$.

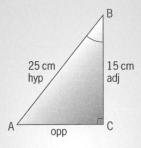

Label the sides and decide on the ratio.

$$\cos\theta = \frac{adj}{hyp}$$

$$\cos\theta = \frac{15}{25}$$

$$\cos\theta = 0.6$$

$$\theta = \cos^{-1} 0.6$$

$$= 53.13°$$

To find the angle you usually use the second function on your calculator.

It is important you know how to use your calculator when working out trigonometry questions.

Progress check

1 Choose a card for each of the missing lengths n on the triangles. The lengths have been rounded to 1 decimal place.

| 12.0 cm | 10.0 cm |
| 12.3 cm | 21.2 cm |

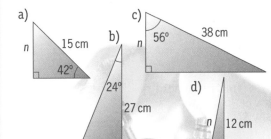

2 Work out the size of the angle x in each of these triangles. Give your answer to the nearest degree.

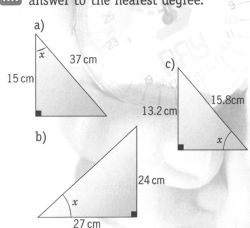

Example 1

The diagram shows a triangle ABC.

Calculate the size of the angle marked $y°$ if

AB = 12.1 cm, CD = 9.7 cm, BÂD = 37°.

Give your answer correct to 1 decimal place.

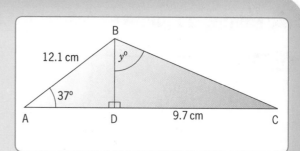

This is an example of a multi-stepped question. In other words you have to do several parts before you get to the final answer.

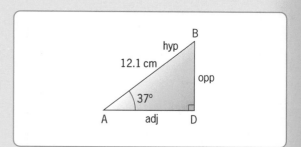

● Find the length of BD first.

$$\sin 37° = \frac{\text{opp}}{\text{hyp}}$$

$$\sin 37° = \frac{BD}{12.1}$$

BD = sin 37° × 12.1

BD = 7.28 cm

BD = 7.3 cm (1dp)

● Now find the angle $y°$.

$$\tan y° = \frac{\text{opp}}{\text{adj}}$$

$$\tan y° = \frac{9.7}{7.3}$$

$$y° = \tan^{-1}(1.3287...)$$

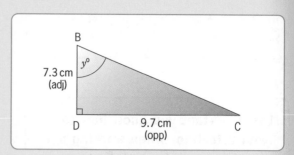

Remember not to round this number off until the end.

$y° = 53°$ (nearest degree)

On the GCSE paper there will always be a question which involves you applying trigonometry or Pythagoras' theorem, or both. Here are some worked examples.

15 MINS

Example 2

A ship sails 37 km due north and then 42 km due east.

Calculate:

a)   the direct distance between the starting point and the finishing point

b)   the bearing of the ship from its starting point

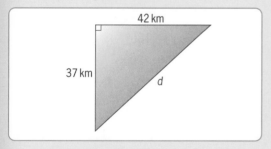

a)   Use Pythagoras' theorem to find the distance (d):

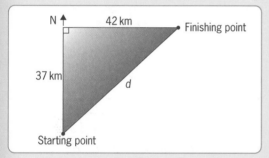

$d^2 = 37^2 + 42^2$

$d^2 = 1369 + 1764$

$d^2 = 3133$

$d = \sqrt{3133}$

$d = 56.0\,km$ (3sf)

b)   Use trigonometry to find the bearing:

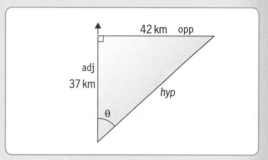

$\tan \theta = \dfrac{opp}{adj}$

$\tan \theta = \dfrac{42}{37}$

$\theta = \tan^{-1}(1.135...)$

$\theta = 48.6°$

Bearing = 049° (nearest degree)

Remember bearings must be 3 figures.

Check that you know these parts of a circle.

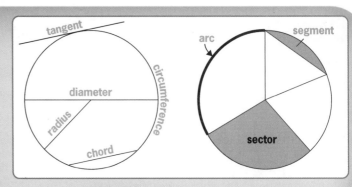

tangent

arc

circumference

diameter

radius

chord

segment

sector

The circle theorems

1. The perpendicular bisector of any chord passes through the centre.

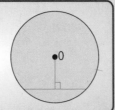

O

2. The angle in a semicircle is always 90°.

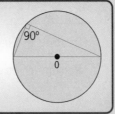

90°

O

3. The radius and a tangent always meet at 90°.

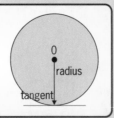

O

radius

tangent

4. Angles in the same segment are equal, e.g. $A\hat{B}C = A\hat{D}C$

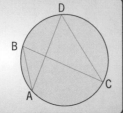

D

B

A

C

5. The angle at the centre is twice the angle at the circumference, e.g. $P\hat{O}Q = 2 \times P\hat{R}Q$

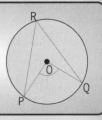

R

O

P

Q

6. Opposite angles of a cyclic quadrilateral add up to 180°.

(A cyclic quadrilateral is a 4-sided shape with each corner touching the circle.)

i.e. $x + y = 180°$

$a + b = 180°$

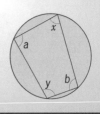

x

a

y

b

7. The lengths of two tangents from a point are equal, e.g. RS = RT

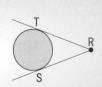

T

R

S

There are several theorems about circles that you need to know.

10 MINS

Examples

Calculate the angles marked *a–d* in the diagram below. Give a reason for your answers.

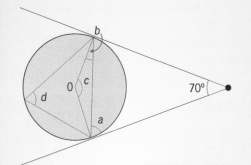

$a = \dfrac{180 - 70°}{2}$	Angles in a triangle add up to 180°.
$= \dfrac{110°}{2}$	Isosceles triangle, base angles are equal.
$= 55°$	
$b = 90° - 55°$	Radius and tangent meet at 90°.
$= 35°$	
$c = 180° - (2 \times 35°)$	Angles in a triangle add up to 180°.
$= 110°$	
$d = 110° \div 2$	Angle at the centre is twice the angle at the circumference.
$= 55°$	

Progress check

Some angles are written on cards. Match the missing angles in the diagrams below with the correct card. O represents the centre of the circle.

53°	50°	62°	109°	126°

a)

b)

c)

d)

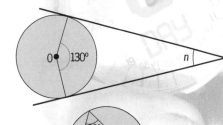

e)

1

2

3

4

5

DAY
6

7

TEST YOURSELF

65

15 MINS

Information to learn

Metric units

Length	Weight	Capacity
10 mm = 1 cm	1000 mg = 1 g	1000 ml = 1 l
100 cm = 1 m	1000 g = 1 kg	100 cl = 1 l
1000 m = 1 m	1000 kg = 1 tonne	1000 cm³ = 1 l
1000 m = 1 km		

Imperial units

Length	Weight	Capacity
1 foot = 12 inches	1 stone = 14 pounds (lb)	20 fluid oz = 1 pint
1 yard = 3 feet	1 pound = 16 ounces (oz)	8 pints = 1 gallon

Comparisons between metric and imperial units

Length	Weight	Capacity
2.5 cm ≈ 1 inch	25 g ≈ 1 ounce	1 litre ≈ $1\frac{3}{4}$ pints
30 cm ≈ 1 foot	1 kg ≈ 2.2 pounds	4.5 litres ≈ 1 gallon
1 m ≈ 39 inches		
8 km ≈ 5 miles		

These comparisons are only approximate.

Compound measures

Speed

Units of speed are: metres per second (m/s)
kilometres per hour (km/h)
miles per hour (mph)

These are known as compound measures because they involve a combination of basic measures.

DAY 6

1 2 3 4 5 7

The metric system of units is based on tens. Older style units called imperial units are not, in general based, on tens.

$$\text{speed (s)} = \frac{\text{distance (d)}}{\text{time (t)}}$$

Rearranging gives:

$$\text{time} = \frac{\text{distance}}{\text{speed}}$$

$$\text{distance} = \text{speed} \times \text{time}$$

Remember to check the units before starting a question. Change them if necessary.

Example

1. A car travels 80 km in 1 hour 20 minutes. Find the speed in km/h.

$$s = \frac{d}{t}, \quad s = \frac{80}{1^{20}/_{60}}$$

$$s = 60 \text{ km/h}$$

Change the time into hours.

20 minutes is $\frac{20}{60}$ of 1 hour.

Density

$$\text{Density} = \frac{\text{mass}}{\text{volume}}, \quad \text{volume} = \frac{\text{mass}}{\text{density}}$$

$$\text{mass} = \text{volume} \times \text{density}$$

Oven cook
Weight: 600g
Temp: 180 °C
Time: 45 seconds per 15 grams

turn over for simpler instructions

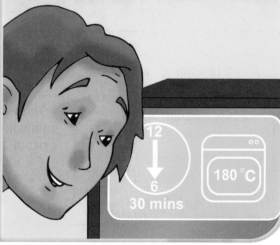

Key formulas you need to learn!

Area of a rectangle

Area = length × width
$A = l \times w$

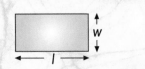

Area of a parallelogram

Area = base × perpendicular height
Remember to use the perpendicular height,
not the slant height. $A = b \times h$

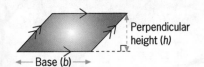
Perpendicular height (h)
Base (b)

Area of a triangle

Area = $\frac{1}{2}$ base × perpendicular height
$A = \frac{1}{2} \times b \times h$

Perpendicular height (h)
Base (b)

Area of a trapezium

You don't need to learn this one as it's on the formula sheet.

Area = $\frac{1}{2}$ × (sum of parallel sides) × perpendicular height between them
$A = \frac{1}{2} \times (a + b) \times h$

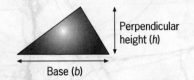
a
Height (h)
b

Circumference of a circle

Circumference = π × diameter
= 2 × π × radius

diameter
0
radius

This is commonly written as:
$C = \pi d$
$C = 2\pi r$

Area of a circle

Area = π × (radius)²
$A = \pi \times r^2$

0 radius

SPEND 15 MINUTES ON THIS TOPIC

The area of a 2-D shape is the amount of flat space that it covers. Common units of area are mm^2, cm^2, m^2.

Example

Find the perimeter and area of this shape.

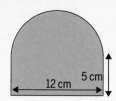

12 cm   5 cm

a) Length of straight sides

$P = 12 + 5 + 5$

$\quad = 22\,cm$

Circumference of semicircle

$C = \dfrac{\pi \times 12}{2}$

$\quad = 18.849...$

Perimeter of shape

$= P + C$

$= 22 + 18.849...$

$= 40.849$

$= 40.8\,cm$ (3sf)

b) Area of rectangle $= l \times w$

$\qquad\qquad\qquad = 12 \times 5$

$\qquad\qquad\qquad = 60\,cm^2$

Area of semicircle $= \dfrac{\pi \times r^2}{2}$

$\qquad\qquad\qquad = \dfrac{\pi \times 6^2}{2}$

$\qquad\qquad\qquad = 56.55\,cm^2$

Total area $= 60 + 56.55$

$\qquad\quad\; = 116.55\,cm^2$

$\qquad\quad\; = 117\,cm^2$ (3sf)

Progress check

1 Find the areas of these shapes:

a)

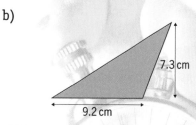

5cm

2cm

10 cm

6 cm

2 cm

8 cm

b)

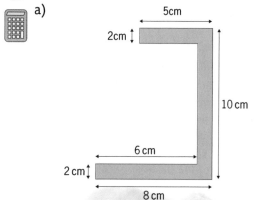

7.3 cm

9.2 cm

c)

6.8 cm

3.7 cm

12.5 cm

2 Find the perimeter and area of this shape:

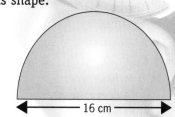

16 cm

Key formulas

Volume of a cuboid

volume = length × width × height
$$V = l \times w \times h$$

To find the surface area of a cuboid, work out the area of each face then add together.
Surface area = $2hl + 2hw + 2lw$

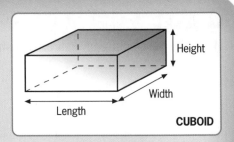

CUBOID

Volume of a prism

volume = area of cross-section × length
$$V = A \times l$$

This formula is given on the formula sheet.

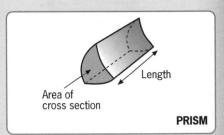

PRISM

Volume of a cylinder

Cylinders are prisms where the cross-section is a circle.
volume = area of cross-section × length

$$V = \pi r^2 \times h$$

area of circle ↗   ↖ height or length

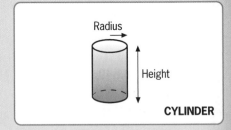

CYLINDER

Examples

Find the volume of this prism:

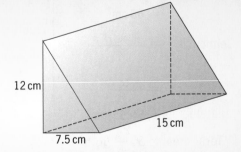

12 cm

15 cm

7.5 cm

$V = A \times l$
$V = (\frac{1}{2} \times b \times h) \times l$
$V = (\frac{1}{2} \times 7.5 \times 12) \times 15$
$V = 675 \, \text{cm}^3$

The area of the cross-section is the area of a triangle.

Cubic units for volume.

SPEND 15 MINUTES ON THIS TOPIC

A prism is any solid that can be cut up into slices that are all the same shape. This is known as having a uniform cross-section.

15 MINS

Example

Find the volume of this cylinder. Leave your answer in terms of π.

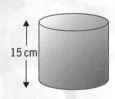

10 cm

←12 cm→

$V = \pi^2 r \times h$
$V = \pi \times 6^2 \times 10$
$V = \pi \times 36 \times 10$
$V = 360\pi \, \text{cm}^3$

Remember to halve the diameter to find the radius.

Example

If the volume of this cylinder is 320 cm³, work out the radius. (Use π = 3.14.) Give your answer to one decimal place.

15 cm

$V = \pi r^2 \times h$

$320 = 3.14 \times r^2 \times 15$ Substitute into the formula.
$320 = 47.1 \, r^2$

$\frac{320}{47.1} = r^2$ Divide both sides by 47.1.

$r = \sqrt{6.79} \ldots$

$r = 2.6 \, \text{cm (1dp)}$ Square root both sides to find the radius.

Progress check

1 The volume of this prism is 1050 cm³.

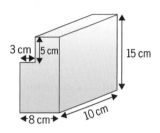

3 cm   5 cm   15 cm

←8 cm→   10 cm

Decide whether this statement is true or false.

2 Work out the volumes of these prisms. Give your answer to 1dp. Use π = 3.142.

a)

2 cm

←6 cm→   8 cm

b)

7.2 cm

←9 cm→   12 cm

c)

10.7 cm

←12.4 cm→

3 If the volume of this cylinder is 205 cm³, work out the height. Use π = 3.14. Give your answer to 3sf.

←5.6 cm→

DAY 6

TEST YOURSELF

Dimensions

> L = length L^2 = area L^3 = volume

- A formula with a mixed dimension is impossible, e.g. $L^2 + L^3$.

- A dimension greater than 3 is impossible.

- Null quantities are numbers, including π and any letters that just stand for numbers, fractions etc.

Examples

If a, b, c represent lengths, what is the dimension of each of these expressions?

1. $\frac{2}{3}a^2 + 4\pi b^2$

   $\rightarrow \frac{2}{3}L^2 + 4\pi L^2$ — Change to dimension letters.

   $\rightarrow L^2 + L^2$ — Remove null quantities.

   $\rightarrow 2L^2$ — Simplify.

   $\rightarrow L^2$ — Remove null quantities.

   (area) — Decide on dimension.

2. $\sqrt{4c^2 + 6a^2}$

   $\rightarrow \sqrt{4L^2 + 6L^2}$

   $\rightarrow \sqrt{L^2 + L^2}$

   $\rightarrow \sqrt{2L^2}$

   $\rightarrow \sqrt{L^2} \rightarrow L$

   (length/perimeter)

Converting units

You need to remember these facts:

Area

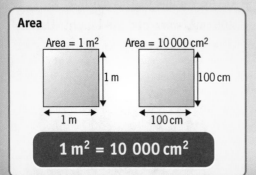

Area = 1 m²   Area = 10 000 cm²

$$1\,m^2 = 10\,000\,cm^2$$

Volume

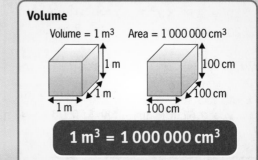

Volume = 1 m³   Area = 1 000 000 cm³

$$1\,m^3 = 1\,000\,000\,cm^3$$

All types of measurement are built up from basic dimensions of length, mass and time.

Remember

When a question has different units, e.g.

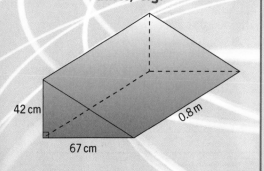

42 cm

0.8 m

67 cm

change the lengths to the same unit before starting the question!

Progress check

1 James says: 'The surface area of a sphere is given by the formula $\frac{4}{3}\pi r^3$.' Explain why this cannot be correct.

2 Decide whether these statements are true or false. The letters v, w and x represent lengths.

a) $\frac{2vwx}{3w^2}$ is a formula for perimeter

b) $\frac{2}{3}\pi \sqrt{v^2 + 2x^2}$ is a formula for area

c) $5\pi \frac{v^2 x^2}{w}$ is a formula for volume

3 Change $2\,m^2$ into cm^2.

4 Change $600\,000\,cm^3$ into m^3.

Drawing a pie chart

When drawing a pie chart you need to:

1. Calculate the angles	2. Draw the pie chart accurately
• Find the total for the items listed	• You are only allowed to be at most 2° out!
• Work out how many degrees one item represents	• Label the sectors
• Work out the degrees for each category	

Example

The favourite subjects of 24 students are shown in the table. Draw a pie chart for this information.

Subject	Frequency	Angle
Maths	9	135°
English	4	60°
Art	5	75°
Geography	6	90°
	24	360°

24 students = 360°

$$1 \text{ student} = \frac{360°}{24} = 15°$$

Now draw the pie chart carefully. The angles must be accurate and each sector must be labelled.

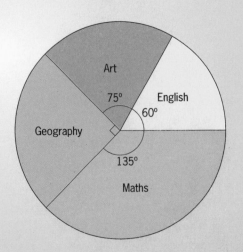

Pie charts are circles split up into sectors.
Each sector represents a certain number of items.

15
MINS

Interpreting a pie chart

When interpreting a pie chart you need to measure the angles carefully if they are not given.

Example

The pie chart shows the favourite sports of 18 students.

How many students like:

a) tennis?

b) football?

c) hockey?

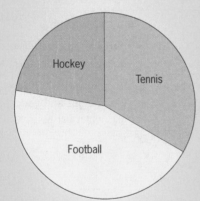

$360° = 18$ students

$1° = \dfrac{18}{360°} = 0.05$ Work out what 1° is worth.

Tennis $= 120° \times 0.05 = 6$ students

Football $= 160° \times 0.05 = 8$ students

Hockey $= 80° \times 0.05 = 4$ students

Progress check

1 Draw a pie chart for this set of data.

Favourite colour	Frequency
Blue	15
Red	9
Black	5
Green	7

2 The pie chart shows the activities that 180 students took part in.

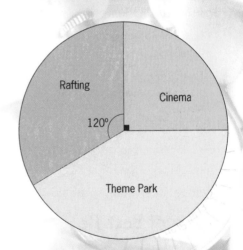

Calculate how many students

a) went rafting

b) went to the cinema

c) went to the Theme Park

SCATTER DIAGRAMS AND CORRELATION

There are three types of correlation

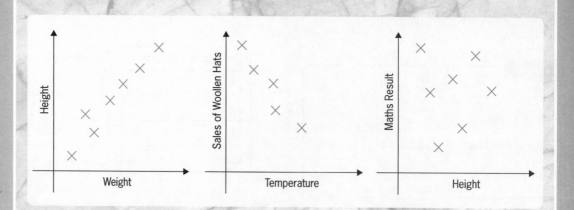

Positive correlation	Negative correlation	Zero correlation
– Both variables are increasing.	– As one variable increases the other decreases.	– Little or no correlation between the variables.
– The taller you are, the more you probably weigh.	– As the temperature increases, the sale of woollen hats probably decreases.	– No connection between your height and your mathematical ability.

Line of best fit

- The line goes as close as possible to all the points.
- There is roughly an equal number of points above the line as below it.

SPEND 10 MINUTES ON THIS TOPIC

Scatter diagrams are used to show two sets of data at the same time. They are important because they show the correlation (connection) between the sets.

10 MINS

Example

The scatter diagram shows the Science and Maths percentages scored by some students.

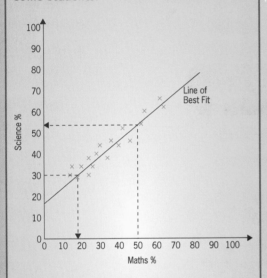

- A student with a Science percentage of 30 would get a Maths percentage of about 17.

- The line of best fit goes in the direction of the data.

- We can estimate that a student with a Maths percentage of 50 would get about 54% in Science.

- This shows how the line of best fit can be used to estimate results.

Progress check

Decide whether these statements are **true** or **false**.

1. There is a positive correlation between the weight of a book and the number of pages.

2. There is no correlation between the height you climb up a mountain and the temperature.

3. There is a negative correlation between the age of a used car and its value.

4. There is a positive correlation between the height of some students and the size of their feet.

5. There is no correlation between the weight of some students and their History GCSE results.

DAY 7

Median – the middle value when the values are put in order of size, e.g.
median of 2 2 3 3 7 9 11 is 3

Mode – the one that occurs the most often, e.g.
mode of 2 2 2 3 5 7 is 2

AVERAGES OF DISCRETE DATA

$\bar{x}$ represents the mean

f represents the frequency

Σ means the sum of

Mean – the most commonly used average

$$\text{Mean} = \frac{\text{sum of a set of values}}{\text{the number of values used}}$$

e.g. mean of 1, 2, 3, 3, 1 is $\dfrac{1 + 2 + 3 + 3 + 1}{5} = 2$

● Finding averages from a frequency table

When the information is in a frequency table finding the averages is a little more difficult.

Example

The table shows the shoe sizes of a group of students.

Shoe size (x)	3	4	5	6	7
Frequency (f)	5	18	22	15	5

Call the shoe size x.

Averages are used to give an idea of a 'typical' value for a set of data. There are three types of average: the mode, median and mean.

a) Finding the mean shoe size

$$\text{mean } (\bar{x}) = \frac{\Sigma fx}{\Sigma f}$$

STEP 1

Multiply the frequency, f, by x
$= (3 \times 5) + (4 \times 18) + (5 \times 22) + (6 \times 15) + (7 \times 5)$

STEP 2

Add up the frequency $= 5 + 18 + 22 + 15 + 5$

STEP 3

$$= \frac{(3 \times 5) + (4 \times 18) + (5 \times 22) + (6 \times 15) + (7 \times 5)}{5 + 18 + 22 + 15 + 5}$$

$$= \frac{322}{65}$$

$$= 4.95 \text{ (2dp)}$$

b) Range of shoe size is $7 - 3$
$= 4$

Range = highest value − lowest value

c) Modal shoe size $= 5$ (highest frequency)

d) Median shoe size: $\frac{\Sigma f + 1}{2}$ tells you how many places along the list you go.
33rd person has size 5 so median $= 5$.

15 MINS

Averages of continuous data

When the data is grouped into class intervals, the exact data is not known.
We estimate the mean by using the midpoints of the class intervals.

Weight (W kg)	Frequency (f)	Midpoint (x)	fx
$30 \leqslant W < 35$	6	32.5	195
$35 \leqslant W < 40$	14	37.5	525
$40 \leqslant W < 45$	22	42.5	935
$45 \leqslant W < 50$	18	47.5	855
	60		2510

$$\bar{x} = \frac{\Sigma fx}{\Sigma f}$$

$$\bar{x} = \frac{2510}{60}$$

$$\bar{x} = 41.83 \text{ (2dp)}$$

Σ means the sum of

f represents the frequency

$\bar{x}$ represents the mean

Add in 2 extra columns – one for
the midpoint and one for fx.

Modal class is $40 \leqslant W < 45$.
This class interval has the highest frequency.

This is a very common question at
GCSE usually worth 4 marks.

Moving averages

- Used to smooth out the changes in a set of data that varies over a period of time.

- A four-point moving average uses four data items in each calculation, a three-point moving average uses three and so on.

Example

Find the four-point moving average for the following data:

3 2 0 1 4 6

Average for 1st 4 data points
$(3 + 2 + 0 + 1) \div 4 = 1.5$

Average for data points 2 to 5
$(2 + 0 + 1 + 4) \div 4 = 1.75$

Average for data points 3 to 6
$(0 + 1 + 4 + 6) \div 4 = 2.75$

- Used to show the trend in a set of data.

- Can be used to draw a trend line on a time series graph.

Progress check

1 Complete the statements for this set of data:

2, 7, 1, 4, 2, 2, 3, 1, 2, 4

a)   the mean of the data is

....

b)   the mode of the data is

....

c)   the range of the data is

....

d)   the median of the data is

....

2 The heights, h cm, of some students are shown in the table.

Height	Frequency
$140 \leqslant h < 145$	4
$145 \leqslant h < 150$	9
$150 \leqslant h < 155$	15
$155 \leqslant h < 160$	6

Calculate an estimate for the mean of this data.

3 Work out the three-point moving average for this data:

2, 1, 3, 4, 5, 6

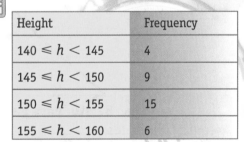

DAY 7

Example

The table shows the marks of 94 students in a Mathematics exam.

a)   Complete the cumulative frequency table for this data.

Mark	Frequency	Mark	Cumulative Frequency	
0–20	2	≤ 20	2	2
21–30	6	≤ 30	8	(2 + 6)
31–40	10	≤ 40	18	(2 + 6 + 10)
41–50	17	≤ 50	35	(2 + 6 + 10 + 17)
51–60	24	≤ 60	59	(2 + 6 + 10 + 17 + 24)
61–70	17	≤ 70	76	(2 + 6 + 10 + 17 + 24 + 17)
71–80	11	≤ 80	87	(2 + 6 + 10 + 17 + 24 + 17 + 11)
81–90	4	≤ 90	91	(2 + 6 + 10 + 17 + 24 + 17 + 11 + 4)
91–100	3	≤ 100	94	(2 + 6 + 10 + 17 + 24 + 17 + 11 + 4 + 3)

This means that 87 students had a score of 80 or less

Cumulative frequency is a running total of all the frequencies.

Cumulative frequency graphs are useful for finding the median and spread of grouped data.

10 MINS

b)   Draw a cumulative frequency graph for this data.

- To do this we must plot the top value of each class interval on the x-axis and the cumulative frequency on the y-axis.

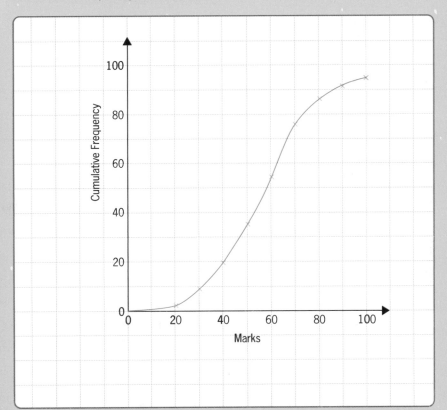

- Plot (20, 2) (30, 8) (40, 18) ...

- Join the points with a smooth curve.

- Since no students had less than zero marks, the graph starts at (0, 0).

With a cumulative frequency graph it is possible to estimate the median of grouped data and the interquartile range.

Upper quartile is three quarters of the way into the distribution
$\frac{3}{4} \times 94 = 70.5$
Read across from 70.5
Upper quartile ≈ 66.5 marks

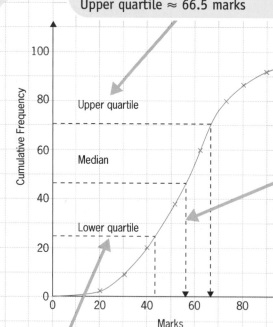

The median splits the data into two halves – the lower 50% and the upper 50%.
Median = $\frac{1}{2} \times$ cumulative frequency
$= \frac{1}{2} \times 94$
$= 47$
Read across from 47 and down to the horizontal axis.
Median ≈ 56 marks

The lower quartile is the value one quarter into the distribution
$\frac{1}{4} \times 94 = 23.5$
Read across from 23.5
Lower quartile ≈ 44 marks

Interquartile range = upper quartile – lower quartile
$= 66.5 - 44$
$= 22.5$ marks

- A large interquartile range indicates that the 'middle half' of the data is widely spread about the median.

- A small interquartile range indicates that the 'middle half' of the data is concentrated about the median.

Box plots

- Sometimes known as box and whisker diagrams.

- Cumulative frequency graphs are not easy to compare; a box plot shows the Interquartile range as a box.

Example

The box plot of the above cumulative frequency graph would look like this:

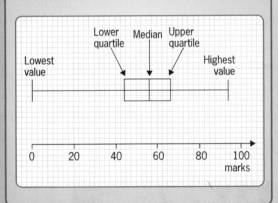

Progress check

The times in minutes to finish an assault course are listed in order.

8, 12, 12, 13, 15, 17,

22, 23, 23, 27, 29

1 Find these:

   a) the lower quartile

   b) the interquartile range

2 Draw a box plot for these data:

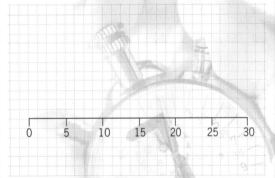

PROBABILITY

Probability of a single event

- **Exhaustive events** account for all possible outcomes.

  e.g. 1, 2, 3, 4, 5, 6 give all possible outcomes when a dice is thrown.

- **Mutually exclusive** events are events that cannot happen at the same time, e.g. a head and a tail on a fair coin cannot appear at the same time.

$$P \text{ (event)} = \frac{\text{number of ways an event can happen}}{\text{total number of outcomes}}$$

- **Theoretical probability** analyses a situation mathematically.

- **Experimental probability** is determined by analysing the results of a number of trials or events.

- This is known as **relative frequency**. e.g. A dice is thrown 55 times. A four comes up 13 times. The relative frequency is $\frac{13}{55}$.

> Probabilities must be written as a fraction, decimal or percentage.

Probability that an event will NOT happen

$$P \text{ (event will not happen)} = 1 - P \text{ (event will happen)}$$

This is the chance or likelihood that something will happen. All probabilities lie between 0 and 1.

Example

The probability that the alarm clock fails to go off is 0.21.

The probability that the alarm clock will go off is
$$1 - 0.21 = 0.79$$

● Expected number

Example

The probability of passing an exam in microbiology is 0.37. If 100 people take the exam. How many are expected to pass.
$$100 \times 0.37 = 37 \text{ people}$$

● Sample space diagrams

These can be helpful when considering the outcomes of two events.

Example

Two spinners are spun and the scores added.

Represent the outcomes on a sample space diagram.

	Spinner 1		
	1	2	3
2	3	4	5
Spinner 2 3	4	5	6
3	4	5	6

Progress check

1. The probability that Ahmed does his homework is 0.65.
   What is the probability that Ahmed does not do his homework?

2. The probability of achieving a grade C in Mathematics is 0.49.
   If 600 students sit this exam, how many would you expect to achieve a grade C?

3. Two fair dice are thrown and their scores are multiplied
   a) Complete the sample space diagram
   b) What is the probability of a score of 6?
   c) What is the probability of an even score?

		Dice 1					
		1	2	3	4	5	6
	1	1	2	3	4	5	6
	2	2			8		
Dice 2	3	3					
	4	4	8			16	
	5	5					
	6	6		18			36

Two rules you need to know first

The OR rule

If two or more events are mutually exclusive the probability of A or B happening is found by adding the probabilities.

> **P (A or B) = P (A) + P (B)**
> **(This also works for more than two outcomes)**

The AND rule

If two or more events are independent, the probability of A and B and C happening together is found by multiplying the separate probabilities

> **P (A and B and C ...) = P (A) x P (B) x P (C) ...**

Example

A bag contains 3 red and 4 blue counters. A counter is taken from the bag at random, its colour is noted and then it is replaced in the bag. A second counter is then taken out of the bag. Draw a tree diagram to illustrate this information.

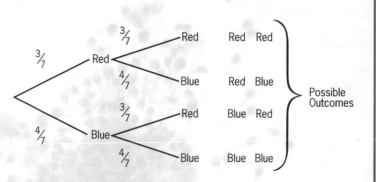

Remember that the branches leaving each point on the tree add up to 1.

These are used to show the possible outcomes of two or more events.

10 MINS

Work out the probability of:

i)   Picking two blues.

ii) One of either colour.

i)   P (two blues) = P (B) × P (B)

$$= \frac{4}{7} \times \frac{4}{7}$$

$$= \frac{16}{49}$$

> **Remember to multiply along the branches.**

ii) P (one of either colour) P (B) × P (R)

$$= \frac{4}{7} \times \frac{3}{7}$$

$$= \frac{12}{49}$$

OR

> **OR means add**

P (R) × P (B)

$$= \frac{3}{7} \times \frac{4}{7}$$

$$= \frac{12}{49}$$

$$= \frac{12}{49} + \frac{12}{49}$$

P (one of either colour)

$$= \frac{24}{49}$$

> **Remember to include all possibilities.**

Progress check

Charlotte has a biased die. The probability of getting a three is 0.4. She rolls the die twice.

1 Complete the tree diagram.

2 Work out the probability that she gets:

a) two threes

b) exactly one three

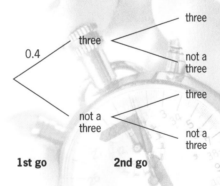

three

0.4

three

not a three

three

not a three

not a three

not a three

1st go   **2nd go**

ANSWERS

Prime factors, HCF and LCM

1. a) $50 = 2 \times 5^2$
   b) $360 = 2^3 \times 3^2 \times 5$
   c) $16 = 2^4$
2. a) False
   b) True
   c) True
   d) False
3. $HCF = 12 \qquad LCM = 144$

Fractions

1. b) $\frac{2}{5} - \frac{5}{7} = \frac{9}{35}$

   c) $\frac{3}{5} \times \frac{1}{11} = \frac{3}{55}$,

   d) $\frac{4}{7} \div 1\frac{2}{3} = \frac{12}{35}$

2. a) $1\frac{8}{21}$
   b) $10\frac{1}{2}$
   c) $2\frac{1}{7}$

Percentage change

1. 51.7% (3sf)
2. 46.7% (3sf)
3. a) 26%
   b) 13%
   c) 19%
   d) 25%

Repeated percentage change and compound interest

1. £5518.28
2. $(1.027)^2 = 1.054729$
3. £121856

Reverse percentage problems

1. a) £52.77
   b) £106.38
   c) £208.51
   d) £446.81
2. a) correct
   b) incorrect
   c) correct

Rounding and estimating

1. answer (d)
2. a) True
   b) False
   c) True
   d) False
3. $\approx \frac{2 \times 800}{0.5} = 3200$
4. 63.5 cm

Indices

1. a) 6^8
   b) 12^{13}
   c) 7^{24}
   d) 5^6
2. a) $6b^{10}$
   b) $2b^{-16}$
   c) $9b^8$
   d) $6b^{14}$

Standard index form

1. a) 6.4×10^4
   b) 2.71×10^5
   c) 4.6×10^{-4}
   d) 7.4×10^{-8}
2. a) 1.2×10^{11}
   b) 2×10^{-1}
   c) 3.5×10^{16}

3. a) 1.4375×10^{18}
   b) 5.48×10^{19}

Formulae and expressions
1. a) $8a - 7b$
   b) $4a^2 - 8b$
   c) $2xy + 2xy^2$
2. a) $\dfrac{-31}{5}$

   b) 4.36 or $\dfrac{109}{25}$

   c) 9
3. $u = \pm \sqrt{v^2 - 2as}$

Brackets and factorisation
1. a) $x^2 + x - 6$
   b) $6x - 8$
   c) $4x^2 - 12x$
   d) $x^2 - 6x + 9$
2. a) $4x(x + 2)$
   b) $6x(2y - x)$
   c) $3ab(a + 2b)$
3. a) $(x + 2)(x^1 + 2)$
   b) $(x - 2)(x - 3)$
   c) $(x + 1)(x - 5)$

Solving linear equations
1. $x = 8$
2. $x = -5$
3. $x = -\frac{1}{2}$
4. $x = -3.25$
5. $x = -\frac{1}{2}$
6. $x = 17$
7. $x = 5.5\,\text{cm}$, shortest length is $2x - 5$
   $= 6\,\text{cm}$

Solving quadratic and cubic equations
1. a) Correct
   b) Incorrect, $x = -5$, $x = 0$
   c) Correct
   d) Correct
2. a) $x = 0$, $x = 7$
   b) $x = -5$, $x = -3$
   c) $x = 2$, $x = 3$
3. $a = 3.3$ (1dp)

Simultaneous equations
1. a) $a = 4$, $b = -4.5$
   b) $p = -3$, $r = 4$
   c) $x = 2$, $y = 3$
2. $x = 2$, $y = 4$

Sequences
1. a) $4n + 1$
   b) $2 - n$
   c) $2n$
   d) $3n + 2$
   e) $5n - 1$
2. a) $3n + 4$
   b) $\dfrac{1}{2n + 1}$
   c) $3n - 2$

Inequalities
1. a) $x < 2.2$
   b) $\dfrac{4}{3} \leqslant x < 3$

   c) $x > \dfrac{-9}{5}$

2

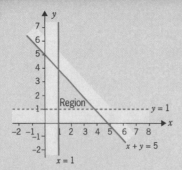

Curved graphs

1 a)

x	−3	−2	−1	0	1	2	3
y	−28	−9	−2	−1	0	7	26

b)

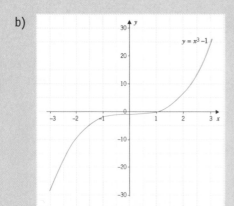

c) $x = 2.5$

2 Graph A: $y = \dfrac{3}{x}$

Graph B: $y = 4x + 2$

Graph C: $y = x^3 - 5$

Graph D: $y = 2 - x^2$

Distance–time graphs

1 False
2 True
3 True
4 True
5 True

Constructions

1 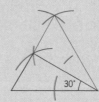 Construct an equilateral triangle first.
Bisect the 60° angle.

2 Perpendicular bisector of an 8 cm line drawn.
3 Angle bisected with compasses only.

Loci

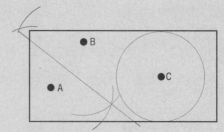

Bearings

1 a) North, clockwise
   b) three
   c) North West
2 a) 072°
   b) 208°
   c) 290°
   d) 139°

Rotations and enlargements

1. Reflection in $y = 0$ (x-axis)
2. Reflection in $x = 0$ (y-axis)
3. Rotation of 180° centre (0, 0)
4. Rotation of 90° anticlockwise. Centre of rotation at (0, 0).
5. Translation of $\binom{5}{-1}$

Pythagoras' theorem

1. Since $26^2 = 24^2 + 10^2$
   $676 = 576 + 100$ and this obeys Pythagoras' theorem, then the triangle must be right-angled.
2. 15.3 cm
3. $\sqrt{149}$

Similarity

1. No
2. a) 20.9 cm
   b) 13.8 cm
   c) 4.5 cm

Trigonometry in right-angled triangles

1. a) 10.0 cm
   b) 12.0 cm
   c) 21.2 cm
   d) 12.3 cm
2. a) 66°
   b) 42°
   c) 57°

Circle theorems

a) 62°
b) 109°
c) 53°
d) 50°
e) 126°

Areas of plane shapes

1. a) 38 cm²
   b) 33.58 cm²
   c) 35.705 cm²
2. Perimeter = 41.13 cm
   Area = 100.53 cm²

Volumes of prisms

1. True
2. a) 96 cm³
   b) 388.8 cm³
   c) 1292.3 cm³
3. 8.33 cm

Dimensions and converting units

1. r^3 represents L^2, which is volume not surface area.
2. a) True
   b) False
   c) True
3. 20 000 cm²
4. 0.6 m³

Pie charts

1. Pie chart drawn with these angles:
   Blue 150°, Red 90°, Black 50°, Green 70°
2. a) Rafting – 60 students
   b) Cinema – 45 students
   c) Theme Park – 75 students

Scatter diagrams and correlation

1. True
2. False
3. True
4. True
5. True

Averages

1. a) 2.8
    b) 2
    c) 6
    d) 2
2   150.88
3   2, 2.6̇, 4, 5

Cumulative frequency graphs and box plots

1   a) lower quartile = 12
    b) interquartile range = 11
2

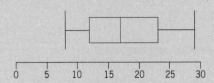

Probability

1   0.35
2   294 students
3
a)

	Dice 1					
	1	2	3	4	5	6
1	1	2	3	4	5	6
2	2	4	6	8	10	12
3	3	6	9	12	15	18
Dice 2 4	4	8	12	16	20	24
5	5	10	15	20	25	30
6	6	12	18	24	30	36

b) 4/36 = 1/9
c) 27/36 = 3/4

Tree diagrams

1

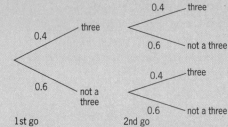

2   a) 0.16
    b) 0.48